British Rai

CW00557767

D
MULTIPLE UNITS

TWENTY-EIGHTH EDITION
2015

The complete guide to all Diesel Multiple Units
and On-Track Machines which operate on the
national railway network and the stock of the
major UK Light Rail & Metro systems

Robert Pritchard & Peter Hall

ISBN 978 1909 431 14 0

© 2014. Platform 5 Publishing Ltd, 52 Broadfield Road, Sheffield, S8 0XJ,
England.

Printed in England by Berforts Information Press, Eynsham, Oxford.

CONTENTS

PROVISION OF INFORMATION

This book has been compiled with care to be as accurate as possible, but in some cases information is not easily available and the publisher cannot be held responsible for any errors or omissions. We would like to thank the companies and individuals which have been co-operative in supplying information to us. The authors of this series of books are always pleased to receive notification from readers of any inaccuracies readers may find in the series, to enhance future editions. Please send comments to:

Robert Pritchard, Platform 5 Publishing Ltd, 52 Broadfield Road, Sheffield, S8 0XJ, England.

e-mail: robert@platform5.com   **Tel:** 0114 255 2625   **Fax:** 0114 255 2471.

This book is updated to information received by 6 October 2014.

UPDATES

This book is updated to the Stock Changes given in **Today's Railways UK 155** (November 2014). Readers are therefore advised to update this book from the official Platform 5 Stock Changes published every month in **Today's Railways UK** magazine, starting with issue 156.

The Platform 5 magazine Today's Railways UK contains news and rolling stock information on the railways of Great Britain and Ireland and is published on the second Monday of every month. For further details of Today's Railways UK, please see the advertisement on the back cover of this book.

Front cover photograph: Arriva Trains-liveried 175 103 approaches Llanelli with the 05.55 Milford Haven–Manchester Piccadilly on 30/06/14.   **Tom McAtee**

BRITAIN'S RAILWAY SYSTEM

INFRASTRUCTURE & OPERATION

Britain's national railway infrastructure is owned by a "not for dividend" company, Network Rail. In September 2014 Network Rail was classified a public sector company, being described by the Government as a "public sector arm's-length body of the Department for Transport".

Many stations and maintenance depots are leased to and operated by Train Operating Companies (TOCs), but some larger stations remain under Network Rail control. The only exception is the infrastructure on the Isle of Wight: Island Line was the only franchise that included the maintenance of the infrastructure as well as the operation of passenger services. As Island Line is now part of the South West Trains franchise, both the infrastructure and trains are operated by South West Trains.

Trains are operated by TOCs over Network Rail, regulated by access agreements between the parties involved. In general, TOCs are responsible for the provision and maintenance of the locomotives, rolling stock and staff necessary for the direct operation of services, whilst Network Rail is responsible for the provision and maintenance of the infrastructure and also for staff to regulate the operation of services.

The Department for Transport is the franchising authority for the national network, with Transport Scotland overseeing the award of the ScotRail franchise. Railway Franchise holders (TOCs) can take commercial risks, although some franchises are known as "management contracts", where ticket revenues pass directly to the DfT. Concessions (such as London Overground) see the operator paid a fee to run the service, usually within tightly specified guidelines. Operators running a Concession would not normally take commercial risks, although there are usually penalties and rewards in the contract.

During 2012 the letting of new franchises was suspended pending a review of the franchise system. The process was restarted in 2013 but it is going to take a number of years to catch-up and several franchises are receiving short-term extensions (or "Direct Awards") in the meantime.

DOMESTIC PASSENGER TRAIN OPERATORS

The large majority of passenger trains are operated by the TOCs on fixed-term franchises. Franchise expiry dates are shown in the list of franchisees below:

Franchise	*Franchisee*	*Trading Name*
Chiltern	Deutsche Bahn (Arriva) (until 31 December 2021)	**Chiltern Railways**
Cross-Country	Deutsche Bahn (Arriva) (until 31 March 2016)*	**CrossCountry**

Franchise extension to be negotiated to November 2019.

| **East Midlands** | Stagecoach Group plc (until 31 March 2015)* | **East Midlands Trains** |

Franchise extension to be negotiated to October 2017.

| **Essex Thameside** | National Express Group plc (until 8 November 2029) | **c2c** |

| **Greater Western** | First Group plc (until 20 September 2015) | **First Great Western** |

Franchise extension currently being negotiated.

| **Greater Anglia** | Abellio (NS) (until 19 October 2016) | **Abellio Greater Anglia** |

| **Integrated Kent** | Govia Ltd (Go-Ahead/Keolis) (until 24 June 2018) | **Southeastern** |

| **InterCity East Coast** | Directly Operated Railways (until 31 March 2015) | **East Coast** |

Currently run on an interim basis by DfT management company Directly Operated Railways (trading as East Coast). This arrangement is due to continue until a new franchise is let to the private sector, with the new franchise currently planned to start on 1 April 2015.

| **InterCity West Coast** | Virgin Rail Group Ltd (until 31 March 2017) | **Virgin Trains** |

| **London Rail** | MTR/Deutsche Bahn (until 12 November 2016) | **London Overground** |

This is a Concession and is different from other rail franchises, as fares and service levels are set by Transport for London instead of the DfT.

| **Merseyrail Electrics** | Serco/Abellio (NS) (until 19 July 2028) | **Merseyrail** |

Under the control of Merseytravel PTE instead of the DfT. Franchise reviewed every five years to fit in with the Merseyside Local Transport Plan.

| **Northern Rail** | Serco/Abellio (NS) (until 1 February 2016) | **Northern** |

| **ScotRail** | First Group plc (until 31 March 2015) | **ScotRail** |

Abellio has won the contract to operate the ScotRail franchise from April 2015.

| **South Central** | Govia Ltd (Go-Ahead/Keolis) (until 25 July 2015) | **Southern** |

Upon termination of the Southern franchise in July 2015 it is to be combined with the new Thameslink, Southern & Great Northern franchise (also operated by Govia).

| **South Western** | Stagecoach Group plc (until 3 February 2017)* | **South West Trains** |

Franchise extension to be negotiated to April 2019.

| **Thameslink & Great Northern** | First Group plc (until 19 September 2021) | **Govia Thameslink Railway** |

| **Trans-Pennine Express** | First Group/Keolis (until 1 April 2015) | **TransPennine Express** |

Franchise extension to be negotiated to February 2016.

Wales & Borders	Deutsche Bahn (Arriva) (until 14 October 2018)*	**Arriva Trains Wales**

The franchise agreement includes the provision for the term to be further extended by mutual agreement by up to five years beyond October 2018. Management of the franchise is devolved to the Welsh Government, but DfT is still the procuring authority.

West Midlands	Govia Ltd (Go-Ahead/Keolis) (until 19 September 2015)*	**London Midland**

Franchise extension to be negotiated to June 2017.

* Franchise agreement includes provision for an extension of up to seven 4-week reporting periods.

The following operators run non-franchised services (* special summer services only):

Operator	Trading Name	Route
BAA	Heathrow Express	London Paddington–Heathrow Airport
First Hull Trains	First Hull Trains	London King's Cross–Hull
Grand Central	Grand Central	London King's Cross–Sunderland/ Bradford Interchange
North Yorkshire Moors Railway Enterprises	North Yorkshire Moors Railway	Pickering–Grosmont–Whitby/ Battersby
West Coast Railway Company	West Coast Railway Company	Birmingham–Stratford-upon-Avon* Fort William–Mallaig* York–Wakefield–York–Scarborough*

INTERNATIONAL PASSENGER OPERATIONS

Eurostar operates passenger services between the UK and mainland Europe.

Eurostar International, established in 2010, is jointly owned by the SNCF (the national operator of France, 55%), SNCB (the national operator of Belgium, 5%) and HM Treasury, 40%). The 40% UK stake was transferred from London & Continental Railways (LCR) to HM Treasury in 2014. LCR had bought British Rail's interest in Eurostar at the time of the UK railway privatisation in 1996.

In addition, a service for the conveyance of accompanied road vehicles through the Channel Tunnel is provided by the tunnel operating company, Eurotunnel.

FREIGHT TRAIN OPERATIONS

The following operators operate freight services or empty passenger stock workings under "Open Access" arrangements:

Colas Rail
DB Schenker Rail (UK)
Devon & Cornwall Railways
Direct Rail Services (DRS)
Freightliner
GB Railfreight (owned by Eurotunnel)
West Coast Railway Company

INTRODUCTION

CLASSIFICATION

Principal details and dimensions are quoted for each class in metric and/or imperial units as considered appropriate bearing in mind common usage in the UK.

All dimensions and weights are quoted for vehicles in an "as new" condition with all necessary supplies (eg oil, water, sand) on board. Dimensions are quoted in the order Length – Width. All lengths quoted are over buffers or couplers as appropriate. Where two lengths are quoted, the first refers to outer vehicles in a set and the second to inner vehicles. All width dimensions quoted are maxima. All weights are shown as metric tonnes (t = tonnes).

LAYOUT OF INFORMATION

DMUs are listed in numerical order of set – using current numbers as allocated by the RSL. Individual "loose" vehicles are listed in numerical order after vehicles formed into fixed formations. Where sets or vehicles have been renumbered in recent years, former numbering detail is shown in parentheses. Each entry is laid out as in the following example:

RSL Set No.	Detail	Livery	Owner	Operator	Depot	Formation	Name
153 309	cr	**GA**	P	*GA*	NC	52309	GERALD FIENNES

Detail Differences. Detail differences which currently affect the areas and types of train which vehicles may work are shown, plus differences in interior layout. Where such differences occur within a class, these are shown either in the heading information or alongside the individual set or vehicle number. The following standard abbreviations are used:

e   European Railway Traffic Management System (ERTMS) signalling equipment fitted.
r   Radio Electric Token Block signalling equipment fitted.

Use of the above abbreviations indicates the equipment fitted is normally operable. Meaning of non-standard abbreviations is detailed in individual class headings.

Set Formations. Regular set formations are shown where these are normally maintained. Readers should note set formations might be temporarily varied from time to time to suit maintenance and/or operational requirements. Vehicles shown as "Spare" are not formed in any regular set formation.

Codes. Codes are used to denote the livery, owner, operation and depot of each unit. Details of these will be found in section 6 of this book. Where a unit or spare car is off-lease, the operator column is left blank.

Names. Only names carried with official sanction are listed. Names are shown in UPPER/lower case characters as actually shown on the name carried on the vehicle(s). Unless otherwise shown, complete units are regarded as named rather than just the individual car(s) which carry the name.

GENERAL INFORMATION

NUMBERING

DMU Classes are listed in class number order.

First generation ("Heritage") DMUs were classified in the series 100–139.
Parry People Movers (not technically DMUs) are classified in the series 139.
Second generation DMUs are classified in the series 140–199.
Diesel Electric Multiple Units are classified in the series 200–249.
Service units are classified in the series 930–999.

First and second generation individual cars are numbered in the series 50000–59999 and 79000–79999.

Parry People Mover cars are numbered in the 39000 series.

DEMU individual cars are numbered in the series 60000–60999, except for a few former EMU vehicles which retain their EMU numbers.

Service Stock individual cars are numbered in the series 975000–975999 and 977000–977999, although this series is not exclusively used for DMU vehicles.

OPERATING CODES

These codes are used by train operating company staff to describe the various different types of vehicles and normally appear on data panels on the inner (ie non driving) ends of vehicles.

The first part of the code describes whether or not the car has a motor or a driving cab as follows:

DM   Driving motor.
M   Motor
DT   Driving trailer
T   Trailer

The next letter is a "B" for cars with a brake compartment.

This is followed by the saloon details:

F   First
S   Standard
C   Composite
so   denotes a semi-open vehicle (part compartments, part open). All other vehicles are assumed to consist solely of open saloons.
L   denotes a vehicle with a toilet.
W   denotes a Wheelchair space.

Finally vehicles with a buffet or kitchen area are suffixed RB or RMB for a miniature buffet counter.

Where two vehicles of the same type are formed within the same unit, the above codes may be suffixed by (A) and (B) to differentiate between the vehicles.

A composite is a vehicle containing both First and Standard Class accommodation, whilst a brake vehicle is a vehicle containing separate specific accommodation for the conductor.

Where vehicles have been declassified, the correct operating code which describes the actual vehicle layout is quoted in this publication.

BUILD DETAILS

Lot Numbers
Vehicles ordered under the auspices of BR were allocated a lot (batch) number when ordered and these are quoted in class headings and sub-headings.

ACCOMMODATION

The information given in class headings and sub-headings is in the form F/S nT (or TD) nW. For example 12/54 1T 1W denotes 12 First Class and 54 Standard Class seats, one toilet and one space for a wheelchair. A number in brackets (ie (2)) denotes tip-up seats (in addition to the fixed seats). Tip-up seats in vestibules do not count. The seating layout of open saloons is shown as 2+1, 2+2 or 3+2 as the case may be. Where units have First Class accommodation as well as Standard Class and the layout is different for each class then these are shown separately prefixed by "1:" and "2:". TD denotes a toilet suitable for use by a disabled person.

ABBREVIATIONS

The following abbreviations are used in class headings and also throughout this publication:

BR	British Railways.
BSI	Bergische Stahl Industrie.
DEMU	Diesel Electric Multiple Unit.
DMU	Diesel Multiple Unit (general term).
EMU	Electric Multiple Unit.
kN	kilonewtons.
km/h	kilometres per hour.
kW	kilowatts.
LT	London Transport.
LUL	London Underground Limited.
m	metres.
mph	miles per hour.
t	tonnes.

1. DIESEL MECHANICAL & DIESEL HYDRAULIC UNITS

1.1 FIRST GENERATION UNITS

CLASS 121   PRESSED STEEL SUBURBAN

First generation units. One set is used on weekdays by Chiltern Railways on peak-hour Aylesbury–Princes Risborough services.
Construction: Steel.
Engines: Two Leyland 1595 of 112 kW (150 hp) at 1800 rpm.
Transmission: Mechanical. Cardan shaft and freewheel to a four-speed epicyclic gearbox and final drive.
Bogies: DD10.
Brakes: Vacuum.
Couplers: Screw.
Dimensions: 20.45 x 2.82 m.
Gangways: Non gangwayed single cars with cabs at each end.
Wheel arrangement: 1-A + A-1.
Doors: Manually-operated slam.
Maximum Speed: 70 mph.
Seating Layout: 3+2 facing.
Multiple Working: "Blue Square" coupling code. First Generation vehicles cannot be coupled to Second Generation units.

Fitted with central door locking.

121 020 formerly in departmental use as unit 960 002 (977722).

121 034 returned to service with Chiltern Railways in 2011. Formerly in departmental use as 977828.

Non-standard livery: 121 020 All over Chiltern blue with a silver stripe.

DMBS. Lot No. 30518 1960–61. –/65. 38.0 t.

121 020	**0**	CR	*CR*	AL	55020
121 034	**G**	CR	*CR*	AL	55034

1.2 PARRY PEOPLE MOVERS

CLASS 139   PPM-60

Gas/flywheel hybrid drive Railcars used on the Stourbridge Junction–
Stourbridge Town branch.
Body construction: Stainless steel framework.
Chassis construction: Welded mild steel box section.
Primary Drive: Ford MVH420 2.3 litre 64 kW (86 hp) LPG fuel engine driving
through Newage marine gearbox, Tandler bevel box and 4 "V" belt driver
to flywheel.
Flywheel Energy Store: 500 kg, 1 m diameter, normal operational speed
range 1000–1500 rpm.
Final transmission: 4 "V" belt driver from flywheel to Tandler bevel box,
Linde hydrostatic transmission and spiral bevel gearbox at No. 2 end axle.
Braking: Normal service braking by regeneration to flywheel (1 m/s/s);
emergency/parking braking by sprung-on, air-off disc brakes (3 m/s/s).
Maximum Speed: 45 mph.
Dimensions: 8.7 x 2.4 m.
Doors: Deans powered doors, double-leaf folding (one per side).
Seating Layout: 1+1 unidirectional/facing.
Multiple Working: Not applicable.

39001–002. DMS. Main Road Sheet Metal, Leyland 2007–08.ₛ–/17(4) 1W. 12.5 t.

139 001	**LM**	P	*LM*	SJ	39001
139 002	**LM**	P	*LM*	SJ	39002

1.3 SECOND GENERATION UNITS

All units in this section have air brakes and are equipped with public
address, with transmission equipment on driving vehicles and flexible
diaphragm gangways. Except where otherwise stated, transmission is
Voith 211r hydraulic with a cardan shaft to a Gmeinder GM190 final drive.

CLASS 142   PACER   BREL DERBY/LEYLAND

DMS–DMSL.

Construction: Steel underframe, rivetted steel body and roof. Built from
Leyland National bus parts on Leyland Bus four-wheeled underframes.
Engines: One Cummins LT10-R of 165 kW (225 hp) at 1950 rpm.
Couplers: BSI at outer ends, bar within unit.
Dimensions: 15.55 x 2.80 m.
Gangways: Within unit only.   **Wheel Arrangement:** 1-A + A-1.
Doors: Twin-leaf inward pivoting.   **Maximum Speed:** 75 mph.
Seating Layout: 3+2 mainly unidirectional bus/bench style unless stated.
Multiple Working: Within class and with Classes 143, 144, 150, 153, 155,
156, 158 and 159.

c Refurbished Arriva Trains Wales units. Fitted with 2+2 individual Chapman seating.
s Fitted with 2+2 individual high-back seating.
t Former First North Western facelifted units – DMS fitted with a luggage/bicycle rack and wheelchair space.
u Merseytravel units – Fitted with 3+2 individual low-back seating.

55542–591. DMS. Lot No. 31003 1985–86. –/62 (c –/46(6) 2W, s –/56, t –/53 or 55 1W, u –/52 or 54 1W). 24.5 t.
55592–641. DMSL. Lot No. 31004 1985–86. –/59 1T (c –/44(6) 1T 2W, s –/50 1T, u –/60 1T). 25.0 t.
55701–746. DMS. Lot No. 31013 1986–87. –/62 (c –/46(6) 2W, s –/56, t –/53 or 55 1W, u –/52 or 54 1W). 24.5 t.
55747–792. DMSL. Lot No. 31014 1986–87. –/59 1T (c –/44(6) 1T 2W, s –/50 1T, u –/60 1T). 25.0 t.

142 001	t	**NO**	A	*NO*	NH	55542	55592
142 002	c	**AV**	A	*AW*	CF	55543	55593
142 003		**NO**	A	*NO*	NH	55544	55594
142 004	t	**NO**	A	*NO*	NH	55545	55595
142 005	t	**NO**	A	*NO*	NH	55546	55596
142 006	c	**AV**	A	*AW*	CF	55547	55597
142 007	t	**NO**	A	*NO*	NH	55548	55598
142 009	t	**NO**	A	*NO*	HT	55550	55600
142 010	c	**AV**	A	*AW*	CF	55551	55601
142 011	t	**NO**	A	*NO*	NH	55552	55602
142 012	t	**NO**	A	*NO*	NH	55553	55603
142 013		**NO**	A	*NO*	NH	55554	55604
142 014	t	**NO**	A	*NO*	NH	55555	55605
142 015	s	**NO**	A	*NO*	HT	55556	55606
142 016	s	**NO**	A	*NO*	HT	55557	55607
142 017	s	**NO**	A	*NO*	HT	55558	55608
142 018	s	**NO**	A	*NO*	HT	55559	55609
142 019	s	**NO**	A	*NO*	HT	55560	55610
142 020	s	**NO**	A	*NO*	HT	55561	55611
142 021	s	**NO**	A	*NO*	HT	55562	55612
142 022	s	**NO**	A	*NO*	HT	55563	55613
142 023	t	**NO**	A	*NO*	HT	55564	55614
142 024	s	**NO**	A	*NO*	HT	55565	55615
142 025	s	**NO**	A	*NO*	HT	55566	55616
142 026	s	**NO**	A	*NO*	HT	55567	55617
142 027	t	**NO**	A	*NO*	HT	55568	55618
142 028	t	**NO**	A	*NO*	NH	55569	55619
142 029		**NO**	A	*NO*	HT	55570	55620
142 030		**NO**	A	*NO*	NH	55571	55621
142 031	t	**NO**	A	*NO*	NH	55572	55622
142 032	t	**NO**	A	*NO*	NH	55573	55623
142 033	t	**NO**	A	*NO*	NH	55574	55624
142 034	t	**NO**	A	*NO*	NH	55575	55625
142 035	t	**NO**	A	*NO*	NH	55576	55626
142 036	t	**NO**	A	*NO*	NH	55577	55627
142 037	t	**NO**	A	*NO*	NH	55578	55628

142 038	t	**NO**	A	*NO*	NH	55579	55629
142 039	t	**NO**	A	*NO*	NH	55580	55630
142 040	t	**NO**	A	*NO*	NH	55581	55631
142 041	u	**NO**	A	*NO*	NH	55582	55632
142 042	u	**NO**	A	*NO*	NH	55583	55633
142 043	u	**NO**	A	*NO*	NH	55584	55634
142 044	u	**NO**	A	*NO*	NH	55585	55635
142 045	u	**NO**	A	*NO*	NH	55586	55636
142 046	u	**NO**	A	*NO*	NH	55587	55637
142 047	u	**NO**	A	*NO*	NH	55588	55638
142 048	u	**NO**	A	*NO*	NH	55589	55639
142 049	u	**NO**	A	*NO*	NH	55590	55640
142 050	s	**NO**	A	*NO*	HT	55591	55641
142 051	u	**NO**	A	*NO*	NH	55701	55747
142 052	u	**NO**	A	*NO*	NH	55702	55748
142 053	u	**NO**	A	*NO*	NH	55703	55749
142 054	u	**NO**	A	*NO*	NH	55704	55750
142 055	u	**NO**	A	*NO*	NH	55705	55751
142 056	u	**NO**	A	*NO*	NH	55706	55752
142 057	u	**NO**	A	*NO*	NH	55707	55753
142 058	u	**NO**	A	*NO*	NH	55708	55754
142 060	t	**NO**	A	*NO*	NH	55710	55756
142 061	t	**NO**	A	*NO*	NH	55711	55757
142 062	t	**NO**	A	*NO*	NH	55712	55758
142 063	t	**NO**	A	*NO*	NH	55713	55759
142 064	t	**NO**	A	*NO*	HT	55714	55760
142 065	s	**NO**	A	*NO*	HT	55715	55761
142 066	s	**NO**	A	*NO*	HT	55716	55762
142 067		**NO**	A	*NO*	HT	55717	55763
142 068	t	**NO**	A	*NO*	HT	55718	55764
142 069	c	**AV**	A	*AW*	CF	55719	55765
142 070	t	**NO**	A	*NO*	HT	55720	55766
142 071	s	**NO**	A	*NO*	HT	55721	55767
142 072	c	**AV**	A	*AW*	CF	55722	55768
142 073	c	**AV**	A	*AW*	CF	55723	55769
142 074	c	**AV**	A	*AW*	CF	55724	55770
142 075	c	**AV**	A	*AW*	CF	55725	55771
142 076	c	**AV**	A	*AW*	CF	55726	55772
142 077	c	**AV**	A	*AW*	CF	55727	55773
142 078	s	**NO**	A	*NO*	HT	55728	55774
142 079	s	**NO**	A	*NO*	HT	55729	55775
142 080	c	**AV**	A	*AW*	CF	55730	55776
142 081	c	**AV**	A	*AW*	CF	55731	55777
142 082	c	**AV**	A	*AW*	CF	55732	55778
142 083	c	**AV**	A	*AW*	CF	55733	55779
142 084	s	**NO**	A	*NO*	HT	55734	55780
142 085	c	**AV**	A	*AW*	CF	55735	55781
142 086	s	**NO**	A	*NO*	HT	55736	55782
142 087	s	**NO**	A	*NO*	HT	55737	55783
142 088	s	**NO**	A	*NO*	HT	55738	55784
142 089	s	**NO**	A	*NO*	HT	55739	55785

142 090	s	**NO**	A	*NO*	HT	55740	55786
142 091	s	**NO**	A	*NO*	HT	55741	55787
142 092	s	**NO**	A	*NO*	HT	55742	55788
142 093	s	**NO**	A	*NO*	HT	55743	55789
142 094	s	**NO**	A	*NO*	HT	55744	55790
142 095	s	**NO**	A	*NO*	HT	55745	55791
142 096	s	**NO**	A	*NO*	HT	55746	55792

CLASS 143   PACER   ALEXANDER/BARCLAY

DMS–DMSL. Similar design to Class 142, but bodies built by W Alexander with Barclay underframes.

Construction: Steel underframe, aluminium alloy body and roof. Alexander bus bodywork on four-wheeled underframes.
Engines: One Cummins LT10-R of 165 kW (225 hp) at 1950 rpm.
Couplers: BSI at outer ends, bar within unit.
Dimensions: 15.45 x 2.80 m.
Gangways: Within unit only.   **Wheel Arrangement:** 1-A + A-1.
Doors: Twin-leaf inward pivoting.   **Maximum Speed:** 75 mph.
Seating Layout: 2+2 high-back Chapman seating, mainly unidirectional.
Multiple Working: Within class and with Classes 142, 144, 150, 153, 155, 156, 158 and 159.

DMS. Lot No. 31005 Andrew Barclay 1985–86. –/48(6) 2W. 24.0 t.
DMSL. Lot No. 31006 Andrew Barclay 1985–86. –/44(6) 1T 2W. 24.5 t.

143 601	**AW**	MG	*AW*	CF	55642	55667	
143 602	**AV**	P	*AW*	CF	55651	55668	
143 603	**FI**	P	*GW*	EX	55658	55669	
143 604	**AV**	P	*AW*	CF	55645	55670	
143 605	**AW**	P	*AW*	CF	55646	55671	
143 606	**AV**	P	*AW*	CF	55647	55672	
143 607	**AV**	P	*AW*	CF	55648	55673	
143 608	**AW**	P	*AW*	CF	55649	55674	
143 609	**AV**	SG	*AW*	CF	55650	55675	Sir Tom Jones
143 610	**AV**	MG	*AW*	CF	55643	55676	
143 611	**FI**	P	*GW*	EX	55652	55677	
143 612	**FI**	P	*GW*	EX	55653	55678	
143 614	**AV**	MG	*AW*	CF	55655	55680	
143 616	**AV**	P	*AW*	CF	55657	55682	
143 617	**FI**	FW	*GW*	EX	55644	55683	
143 618	**FI**	FW	*GW*	EX	55659	55684	
143 619	**FI**	FW	*GW*	EX	55660	55685	
143 620	**FI**	P	*GW*	EX	55661	55686	
143 621	**FI**	P	*GW*	EX	55662	55687	
143 622	**AV**	P	*AW*	CF	55663	55688	
143 623	**AV**	P	*AW*	CF	55664	55689	
143 624	**AW**	P	*AW*	CF	55665	55690	
143 625	**AV**	P	*AW*	CF	55666	55691	

CLASS 144   PACER   ALEXANDER/BREL DERBY

DMS–DMSL or DMS–MS–DMSL. As Class 143, but underframes built by BREL.

Construction: Steel underframe, aluminium alloy body and roof. Alexander bus bodywork on four-wheeled underframes.
Engines: One Cummins LT10-R of 165 kW (225 hp) at 1950 rpm.
Couplers: BSI at outer ends, bar within unit.
Dimensions: 15.45/15.43 x 2.80 m.
Gangways: Within unit only.   **Wheel Arrangement:** 1-A + A-1.
Doors: Twin-leaf inward pivoting.   **Maximum Speed:** 75 mph.
Seating Layout: 2+2 high-back Richmond seating, mainly unidirectional.
Multiple Working: Within class and with Classes 142, 143, 150, 153, 155, 156, 158 and 159.

DMS. Lot No. 31015 BREL Derby 1986–87. –/45(3) 1W 24.0 t.
MS. Lot No. BREL Derby 31037 1987. –/58. 23.5 t.
DMSL. Lot No. BREL Derby 31016 1986–87. –/41(3) 1T. 24.5 t.

144 001	**NO**	P	*NO*	NL	55801		55824
144 002	**NO**	P	*NO*	NL	55802		55825
144 003	**NO**	P	*NO*	NL	55803		55826
144 004	**NO**	P	*NO*	NL	55804		55827
144 005	**NO**	P	*NO*	NL	55805		55828
144 006	**NO**	P	*NO*	NL	55806		55829
144 007	**NO**	P	*NO*	NL	55807		55830
144 008	**NO**	P	*NO*	NL	55808		55831
144 009	**NO**	P	*NO*	NL	55809		55832
144 010	**NO**	P	*NO*	NL	55810		55833
144 011	**NO**	P	*NO*	NL	55811		55834
144 012	**NO**	P	*NO*	NL	55812		55835
144 013	**NO**	P	*NO*	NL	55813		55836
144 014	**NO**	P	*NO*	NL	55814	55850	55837
144 015	**NO**	P	*NO*	NL	55815	55851	55838
144 016	**NO**	P	*NO*	NL	55816	55852	55839
144 017	**NO**	P	*NO*	NL	55817	55853	55840
144 018	**NO**	P	*NO*	NL	55818	55854	55841
144 019	**NO**	P	*NO*	NL	55819	55855	55842
144 020	**NO**	P	*NO*	NL	55820	55856	55843
144 021	**NO**	P	*NO*	NL	55821	55857	55844
144 022	**NO**	P	*NO*	NL	55822	55858	55845
144 023	**NO**	P	*NO*	NL	55823	55859	55846

Name: 144 001 THE PENISTONE LINE PARTNERSHIP

CLASS 150/0   SPRINTER   BREL YORK

DMSL–MS–DMS. Prototype Sprinter.

Construction: Steel.
Engines: One Cummins NT855R5 of 213 kW (285 hp) at 2100 rpm.
Bogies: BX8P (powered), BX8T (non-powered).
Couplers: BSI at outer end of driving vehicles, bar non-driving ends.

Dimensions: 19.93/19.92 x 2.73 m.
Gangways: Within unit only.   **Wheel Arrangement:** 2-B + 2-B + B-2.
Doors: Twin-leaf sliding.   **Maximum Speed:** 75 mph.
Seating Layout: 3+2 (mainly unidirectional).
Multiple Working: Within class and with Classes 142, 143, 144, 153, 155, 156, 158, 159, 170 and 172.

DMSL. Lot No. 30984 1984. –/72 1T. 35.4 t.
MS. Lot No. 30986 1984. –/92. 35.0 t.
DMS. Lot No. 30985 1984. –/69(6). 34.7 t.

| 150 001 | | **FB** | A | *GW* | RG | 55200 | 55400 | 55300 |
| 150 002 | | **FB** | A | *GW* | RG | 55201 | 55401 | 55301 |

CLASS 150/1   SPRINTER   BREL YORK

DMSL–DMS.

Construction: Steel.
Engines: One Cummins NT855R5 of 213 kW (285 hp) at 2100 rpm.
Bogies: BP38 (powered), BT38 (non-powered).
Couplers: BSI.
Dimensions: 19.74 x 2.82 m.
Gangways: Within unit only.   **Wheel Arrangement:** 2-B (+ 2-B) + B-2.
Doors: Twin-leaf sliding.   **Maximum Speed:** 75 mph.
Seating Layout: 3+2 facing as built but Centro units were reseated with mainly unidirectional seating.
Multiple Working: Within class and with Classes 142, 143, 144, 153, 155, 156, 158, 159, 170 and 172.

c 3+2 Chapman seating.

DMSL. Lot No. 31011 1985–86. –/72 1T (c –/59 1TD (except 52144 which is –/62 1TD), t –/71 1T, u –/71 1T). 38.3 t.
DMS. Lot No. 31012 1985–86. –/76 (c –/65, t –/73, u –/70 (6)). 38.1 t.

150 101	u	**FB**	A	*GW*	PM	52101	57101
150 102	u	**FB**	A	*GW*	PM	52102	57102
150 103	u	**NO**	A	*NO*	NH	52103	57103
150 104	u	**FB**	A	*GW*	PM	52104	57104
150 105	u	**LM**	A	*LM*	TS	52105	57105
150 106	u	**FB**	A	*GW*	PM	52106	57106
150 107	u	**LM**	A	*LM*	TS	52107	57107
150 108	u	**FB**	A	*GW*	PM	52108	57108
150 109	u	**LM**	A	*LM*	TS	52109	57109
150 110	u	**NO**	A	*NO*	NH	52110	57110
150 111	u	**NO**	A	*NO*	NH	52111	57111
150 112	u	**NO**	A	*NO*	NH	52112	57112
150 113	u	**NO**	A	*NO*	NH	52113	57113
150 114	u	**NO**	A	*NO*	NH	52114	57114
150 115	u	**NO**	A	*NO*	NH	52115	57115
150 116	u	**NO**	A	*NO*	NH	52116	57116
150 117	u	**NO**	A	*NO*	NH	52117	57117
150 118	u	**NO**	A	*NO*	NH	52118	57118

▲ One of the two Class 121 "bubble cars" used on the Princes Risborough–Aylesbury line, BR green-liveried 121 034, approaches Little Kimble with the 11.11 Princes Risborough–Aylesbury on 25/05/11. **Robert Pritchard**

▼ London Midland-liveried Parry People Mover 139 002 leaves Stourbridge Town with the 10.35 shuttle to Stourbridge Junction on 21/09/14. **Robert Pritchard**

▲ Arriva Trains liveried 142 073 and 143 623 arrive at Radyr with the 10.47 Treherbert–Cardiff Central on 17/10/12. **Robert Pritchard**

▼ First Great Western "Local Lines"-liveried 143 603 and 153 373 at Exeter St Davids with the 17.50 to Exmouth on 06/04/13. **Robert Pritchard**

▲ Northern-liveried 144 021 heads away from Morecambe with the 10.19 Leeds–Heysham Port on 15/06/13. **Dave McAlone**

▼ 150 133 and an unidentified 156 arrive at Preston with the 18.20 Manchester Victoria–Blackpool North on 08/06/13. **Robin Ralston**

15

▲ London Midland-liveried 153 366 awaits departure from Bletchley with the 11.05 to Bedford on 11/04/14. **Harry Savage**

▼ Northern-liveried 155 345 passes Hambleton West Junction with the 12.54 Selby–Huddersfield on 05/08/14. **Lindsay Atkinson**

▲ ScotRail Saltire-liveried 156 514, the last of the 114-strong class to be built, leaves Glasgow Central empty stock on 02/06/13. **Robert Pritchard**

▼ First Great Western 3-car hybrid 158 960 pauses at Filton Abbey Wood with the 16.35 Cardiff Central–Brighton on 18/08/13. **Robert Pritchard**

▲ South West Trains 159 005 leaves Exeter St Davids with the 14.26 to London Waterloo on 26/07/13. **Stewart Armstrong**

▼ Chiltern Railways 165 012 leaves Bicester North with the 16.48 Banbury–London Marylebone on 25/05/12. **Robert Pritchard**

▲ The first of the FGW Class 166s to receive the all over blue livery, 166 221, passes Old Linslade on the WCML on 12/07/14 on its way back from Wolverton to Reading following overhaul. **Mark Beal**

▼ In the new Chiltern Mainline livery, 168 003 passes Hatton North Jn with the 16.18 London Marylebone–Birmingham Snow Hill on 19/08/14. **Dave Gommersall**

▲ ScotRail Saltire-liveried 170 470 passes Plean with the 14.18 Dunblane–Glasgow Queen Street on 23/08/14. **Ian Lothian**

▼ London Midland 172 215 arrives at Birmingham Moor Street with the 14.29 Stratford-upon-Avon–Worcester Foregate Street on 29/09/13. **Robert Pritchard**

150 119	u	**NO**	A	*NO*	NH	52119	57119
150 120	t	**FB**	A	*GW*	EX	52120	57120
150 121	u	**FB**	A	*GW*	PM	52121	57121
150 122	u	**FB**	A	*GW*	EX	52122	57122
150 123	t	**FB**	A	*GW*	EX	52123	57123
150 124	u	**FB**	A	*GW*	EX	52124	57124
150 127	u	**FB**	A	*GW*	PM	52127	57127
150 128	t	**FB**	A	*GW*	EX	52128	57128
150 129	t	**FB**	A	*GW*	EX	52129	57129
150 130	t	**FB**	A	*GW*	EX	52130	57130
150 131	t	**FB**	A	*GW*	EX	52131	57131
150 132	u	**NO**	A	*NO*	NH	52132	57132
150 133	c	**NO**	A	*NO*	NH	52133	57133
150 134	c	**NO**	A	*NO*	NH	52134	57134
150 135	c	**NO**	A	*NO*	NH	52135	57135
150 136	c	**NO**	A	*NO*	NH	52136	57136
150 137	c	**NO**	A	*NO*	NH	52137	57137
150 138	c	**NO**	A	*NO*	NH	52138	57138
150 139	c	**NO**	A	*NO*	NH	52139	57139
150 140	c	**NO**	A	*NO*	NH	52140	57140
150 141	c	**NO**	A	*NO*	NH	52141	57141
150 142	c	**NO**	A	*NO*	NH	52142	57142
150 143	c	**NO**	A	*NO*	NH	52143	57143
150 144	c	**NO**	A	*NO*	NH	52144	57144
150 145	c	**NO**	A	*NO*	NH	52145	57145
150 146	c	**NO**	A	*NO*	NH	52146	57146
150 147	c	**NO**	A	*NO*	NH	52147	57147
150 148	c	**NO**	A	*NO*	NH	52148	57148
150 149	c	**NO**	A	*NO*	NH	52149	57149
150 150	c	**NO**	A	*NO*	NH	52150	57150

Names:

150 125	THE HEART OF WESSEX LINE
150 129	Devon & Cornwall RAIL PARTNERSHIP
150 130	Severnside Community Rail Partnership

CLASS 150/2   SPRINTER   BREL YORK

DMSL–DMS.

Construction: Steel.
Engines: One Cummins NT855R5 of 213 kW (285 hp) at 2100 rpm.
Bogies: BP38 (powered), BT38 (non-powered).
Couplers: BSI.
Dimensions: 19.74 x 2.82 m.
Gangways: Throughout.   **Wheel Arrangement:** 2-B + B-2.
Doors: Twin-leaf sliding.   **Maximum Speed:** 75 mph.
Seating Layout: 3+2 mainly unidirectional seating as built, but most units have now been refurbished with new 2+2 seating.
Multiple Working: Within class and with Classes 142, 143, 144, 153, 155, 156, 158, 159, 170 and 172.

c 3+2 Chapman seating (former First North Western units).
p Refurbished Arriva Trains Wales units with 2+2 Primarius seating.
v Units refurbished for Valley Lines with 2+2 Chapman seating.
w Units refurbished for First Great Western with 2+2 Chapman seating.

Northern promotional vinyls:

150 203/205/207/215/218/222/223/225/228/268–271/273–277 Welcome
to Yorkshire.
150 272 R&B Festival week, Colne.

DMSL. Lot No. 31017 1986–87. † –/68 1T 1W, c –/62 1TD, p –/60(4) 1T, u –/71
1T), v –/60(8) 1T, w –/60(8) 1T. 37.5 t.
DMS. Lot No. 31018 1986–87. † –/71(3), c –/70, p –/56(10) 1W, u –/70(6),
v –/56(15) 2W, w –/56(17) 2W, z –/68. 36.5 t.

150 201	c	**NO**	A	*NO*	NH	52201	57201
150 202	u	**FB**	A	*GW*	PM	52202	57202
150 203	c	**NO**	A	*NO*	NH	52203	57203
150 204	u	**NO**	A	*NO*	NH	52204	57204
150 205	c	**NO**	A	*NO*	NH	52205	57205
150 206	u	**NO**	A	*NO*	NH	52206	57206
150 207	c	**NO**	A	*NO*	NH	52207	57207
150 208	p	**AV**	P	*AW*	CF	52208	57208
150 210	u	**NO**	A	*NO*	NH	52210	57210
150 211	c	**NO**	A	*NO*	NH	52211	57211
150 213	p	**AW**	P	*AW*	CF	52213	57213
150 214	u	**NO**	A	*NO*	NH	52214	57214
150 215	c	**NO**	A	*NO*	NH	52215	57215
150 216	u	**FB**	A	*GW*	PM	52216	57216
150 217	p	**AV**	P	*AW*	CF	52217	57217
150 218	c	**NO**	A	*NO*	NH	52218	57218
150 219	w	**FI**	P	*GW*	PM	52219	57219
150 220	u	**NO**	A	*NO*	NH	52220	57220
150 221	w	**FI**	P	*GW*	PM	52221	57221
150 222	c	**NO**	A	*NO*	NH	52222	57222
150 223	c	**NO**	A	*NO*	NH	52223	57223
150 224	c	**NO**	A	*NO*	NH	52224	57224
150 225	c	**NO**	A	*NO*	NH	52225	57225
150 226	u	**NO**	A	*NO*	NH	52226	57226
150 227	p	**AW**	P	*AW*	CF	52227	57227
150 228	†	**NO**	P	*NO*	NH	52228	57228
150 229	p	**AV**	P	*AW*	CF	52229	57229
150 230	w	**AW**	P	*AW*	CF	52230	57230
150 231	p	**AV**	P	*AW*	CF	52231	57231
150 232	w	**FI**	P	*GW*	PM	52232	57232
150 233	w	**FI**	P	*GW*	PM	52233	57233
150 234	w	**FI**	P	*GW*	PM	52234	57234
150 235	p	**AV**	P	*AW*	CF	52235	57235
150 236	w	**AW**	P	*AW*	CF	52236	57236
150 237	p	**AW**	P	*AW*	CF	52237	57237
150 238	w	**FI**	P	*GW*	PM	52238	57238
150 239	w	**FI**	P	*GW*	PM	52239	57239

150 240	w	**AV**	P	*AW*	CF	52240	57240
150 241	w	**AV**	P	*AW*	CF	52241	57241
150 242	w	**AV**	P	*AW*	CF	52242	57242
150 243	w	**FI**	P	*GW*	PM	52243	57243
150 244	w	**FI**	P	*GW*	PM	52244	57244
150 245	p	**AV**	P	*AW*	CF	52245	57245
150 246	w	**FI**	P	*GW*	PM	52246	57246
150 247	w	**FI**	P	*GW*	PM	52247	57247
150 248	w	**FI**	P	*GW*	PM	52248	57248
150 249	w	**FI**	P	*GW*	PM	52249	57249
150 250	p	**AW**	P	*AW*	CF	52250	57250
150 251	w	**AW**	P	*AW*	CF	52251	57251
150 252	p	**AV**	P	*AW*	CF	52252	57252
150 253	w	**AW**	P	*AW*	CF	52253	57253
150 254	w	**AV**	P	*AW*	CF	52254	57254
150 255	p	**AW**	P	*AW*	CF	52255	57255
150 256	p	**AV**	P	*AW*	CF	52256	57256
150 257	p	**AW**	P	*AW*	CF	52257	57257
150 258	p	**AV**	P	*AW*	CF	52258	57258
150 259	p	**AV**	P	*AW*	CF	52259	57259
150 260	p	**AV**	P	*AW*	CF	52260	57260
150 261	w	**FB**	P	*GW*	PM	52261	57261
150 262	p	**AV**	P	*AW*	CF	52262	57262
150 263	w	**FI**	P	*GW*	PM	52263	57263
150 264	p	**AV**	P	*AW*	CF	52264	57264
150 265	w	**FI**	P	*GW*	PM	52265	57265
150 266	w	**FI**	P	*GW*	PM	52266	57266
150 267	v	**AV**	P	*AW*	CF	52267	57267
150 268	†	**NO**	P	*NO*	NH	52268	57268
150 269	†	**NO**	P	*NO*	NH	52269	57269
150 270	†	**NO**	P	*NO*	NH	52270	57270
150 271	†	**NO**	P	*NO*	NH	52271	57271
150 272	†	**NO**	P	*NO*	NH	52272	57272
150 273	†	**NO**	P	*NO*	NH	52273	57273
150 274	†	**NO**	P	*NO*	NH	52274	57274
150 275	†	**NO**	P	*NO*	NH	52275	57275
150 276	†	**NO**	P	*NO*	NH	52276	57276
150 277	†	**NO**	P	*NO*	NH	52277	57277
150 278	v	**AW**	P	*AW*	CF	52278	57278
150 279	v	**AW**	P	*AW*	CF	52279	57279
150 280	v	**AV**	P	*AW*	CF	52280	57280
150 281	v	**AW**	P	*AW*	CF	52281	57281
150 282	v	**AV**	P	*AW*	CF	52282	57282
150 283	p	**AV**	P	*AW*	CF	52283	57283
150 284	p	**AW**	P	*AW*	CF	52284	57284
150 285	p	**AV**	P	*AW*	CF	52285	57285

CLASS 150/9   SPRINTER   BREL YORK

3-car First Great Western hybrids formed of a Class 150/1 with a 150/2 centre vehicle. DMSL–DMS–DMS. For details see Class 150/1 or Class 150/2.

150 925	u	**FB**	A	*GW*	PM	52125	57209	57125
150 926	u	**FB**	A	*GW*	PM	52126	57212	57126

CLASS 153   SUPER SPRINTER   LEYLAND BUS

DMSL. Converted by Hunslet-Barclay, Kilmarnock from Class 155 2-car units.

Construction: Steel underframe, rivetted steel body and roof. Built from Leyland National bus parts on Leyland Bus bogied underframes.
Engine: One Cummins NT855R5 of 213 kW (285 hp) at 2100 rpm.
Bogies: One P3-10 (powered) and one BT38 (non-powered).
Couplers: BSI.
Dimensions: 23.21 x 2.70 m.
Gangways: Throughout.   **Wheel Arrangement:** 2-B.
Doors: Single-leaf sliding plug.   **Maximum Speed:** 75 mph.
Seating Layout: 2+2 facing/unidirectional.
Multiple Working: Within class and with Classes 142, 143, 144, 150, 155, 156, 158, 159, 170 and 172.

Cars numbered in the 573xx series were renumbered by adding 50 to their original number so that the last two digits correspond with the set number.

c Chapman seating.
d Richmond seating.

52301–52335. DMSL. Lot No. 31026 1987–88. Converted under Lot No. 31115 1991–92. –/72(3) 1T 1W. (s –/72 1T 1W, t –/72(2) 1T 1W). 41.2 t.
57301–57335. DMSL. Lot No. 31027 1987–88. Converted under Lot No. 31115 1991–92. –/72 1T 1W (s –/72 1T 1W). 41.2 t.

153 301	d	**NO**	A	*NO*	NL	52301	
153 302	c	**EM**	A	*EM*	NM	52302	
153 303	c	**AW**	A	*AW*	CF	52303	
153 304	ds	**NO**	A	*NO*	NL	52304	
153 305	d	**FI**	A	*GW*	EX	52305	
153 306	cr	**GA**	P	*GA*	NC	52306	
153 307	d	**NO**	A	*NO*	NL	52307	
153 308	c	**EM**	A	*EM*	NM	52308	
153 309	cr	**GA**	P	*GA*	NC	52309	GERARD FIENNES
153 310	c	**EM**	P	*EM*	NM	52310	
153 311	c	**EM**	P	*EM*	NM	52311	
153 312	s	**AV**	A	*AW*	CF	52312	
153 313	cs	**EM**	P	*EM*	NM	52313	
153 314	cr	**1**	P	*GA*	NC	52314	
153 315	ds	**NO**	A	*NO*	NL	52315	
153 316	c	**NO**	P	*NO*	NL	52316	John "Logitude" Harrison
							Inventor of the Marine Chronometer
153 317	ds	**NO**	A	*NO*	NL	52317	

153 318	d	**FI**	A	*GW*	EX	52318
153 319	c	**EM**	A	*EM*	NM	52319
153 320	c	**AV**	P	*AW*	CF	52320
153 321	ct	**EM**	P	*EM*	NM	52321
153 322	cr	**GA**	P	*GA*	NC	52322	BENJAMIN BRITTEN
153 323	c	**AV**	P	*AW*	CF	52323
153 324	c	**NO**	P	*NO*	NL	52324
153 325	c	**LM**	P	*GW*	EX	52325
153 326	c	**EM**	P	*EM*	NM	52326
153 327	c	**AV**	A	*AW*	CF	52327
153 328	ds	**NO**	A	*NO*	NL	52328
153 329	c	**FB**	P	*GW*	EX	52329
153 330	cs	**NO**	P	*NO*	NL	52330
153 331	d	**NO**	A	*NO*	NL	52331
153 332	c	**NO**	P	*NO*	NL	52332
153 333	cs	**LM**	P	*GW*	EX	52333
153 334	ct	**LM**	P	*LM*	TS	52334
153 335	cr	**GA**	P	*GA*	NC	52335	MICHAEL PALIN
153 351	d	**NO**	A	*NO*	NL	57351
153 352	ds	**NO**	A	*NO*	NL	57352
153 353	c	**AW**	A	*AW*	CF	57353
153 354	c	**LM**	P	*LM*	TS	57354
153 355	c	**EM**	A	*EM*	NM	57355
153 356	c	**LM**	P	*LM*	TS	57356
153 357	c	**EM**	A	*EM*	NM	57357
153 358	c	**NO**	P	*NO*	NL	57358
153 359	c	**NO**	P	*NO*	NL	57359
153 360	c	**NO**	P	*NO*	NL	57360
153 361	cs	**FB**	P	*GW*	EX	57361
153 362	cs	**AW**	A	*AW*	CF	57362
153 363	cs	**NO**	P	*NO*	NL	57363
153 364	c	**LM**	P	*LM*	TS	57364
153 365	c	**LM**	P	*LM*	TS	57365
153 366	c	**LM**	P	*LM*	TS	57366
153 367	cs	**AV**	P	*AW*	CF	57367
153 368	d	**FI**	A	*GW*	EX	57368
153 369	c	**FB**	P	*GW*	EX	57369
153 370	d	**FI**	A	*GW*	EX	57370
153 371	c	**LM**	P	*LM*	TS	57371
153 372	d	**FI**	A	*GW*	EX	57372
153 373	d	**FI**	A	*GW*	EX	57373
153 374	c	**EM**	A	*EM*	NM	57374
153 375	c	**LM**	P	*LM*	TS	57375
153 376	c	**EM**	P	*EM*	NM	57376	X24-EXPEDITIOUS
153 377	d	**FI**	A	*GW*	EX	57377
153 378	d	**NO**	A	*NO*	NL	57378
153 379	c	**EM**	P	*EM*	NM	57379
153 380	d	**FI**	A	*GW*	EX	57380
153 381	c	**EM**	P	*EM*	NM	57381
153 382	d	**FI**	A	*GW*	EX	57382
153 383	c	**EM**	P	*EM*	NM	57383

| 153 384 | c | **EM** | P | *EM* | NM | 57384 |
| 153 385 | c | **EM** | P | *EM* | NM | 57385 |

CLASS 155   SUPER SPRINTER   LEYLAND BUS

DMSL–DMS.

Construction: Steel underframe, rivetted steel body and roof. Built from Leyland National bus parts on Leyland Bus bogied underframes.
Engines: One Cummins NT855R5 of 213 kW (285 hp) at 2100 rpm.
Bogies: One P3-10 (powered) and one BT38 (non-powered).
Couplers: BSI.
Dimensions: 23.21 x 2.70 m.
Gangways: Throughout.   **Wheel Arrangement:** 2-B + B-2.
Doors: Single-leaf sliding plug.   **Maximum Speed:** 75 mph.
Seating Layout: 2+2 facing/unidirectional Chapman seating.
Multiple Working: Within class and with Classes 142, 143, 144, 150, 153, 156, 158, 159, 170 and 172.

Northern promotional vinyls:

155 341–347 Leeds–Bradford–Manchester route (the "Calder Valley").

DMSL. Lot No. 31057 1988. –/76 1TD 1W. 39.0 t.
DMS. Lot No. 31058 1988. –/80. 38.6 t.

155 341	**NO**	P	*NO*	NL	52341	57341
155 342	**NO**	P	*NO*	NL	52342	57342
155 343	**NO**	P	*NO*	NL	52343	57343
155 344	**NO**	P	*NO*	NL	52344	57344
155 345	**NO**	P	*NO*	NL	52345	57345
155 346	**NO**	P	*NO*	NL	52346	57346
155 347	**NO**	P	*NO*	NL	52347	57347

CLASS 156   SUPER SPRINTER   METRO-CAMMELL

DMSL–DMS.

Construction: Steel.
Engines: One Cummins NT855R5 of 213 kW (285 hp) at 2100 rpm.
Bogies: One P3-10 (powered) and one BT38 (non-powered).
Couplers: BSI.
Dimensions: 23.03 x 2.73 m.
Gangways: Throughout.   **Wheel Arrangement:** 2-B + B-2.
Doors: Single-leaf sliding.   **Maximum Speed:** 75 mph.
Seating Layout: 2+2 facing/unidirectional.
Multiple Working: Within class and with Classes 142, 143, 144, 150, 153, 155, 158, 159, 170 and 172.

† Greater Anglia units fitted with new universal access toilet to meet the 2020 accessibility regulations.
c Chapman seating.
d Richmond seating.

Northern promotional vinyls:

156 441	Manchester and Liverpool
156 461	Ravenglass & Eskdale Railway.
156 464	Lancashire DalesRail
156 484	Settle & Carlisle line.

DMSL. Lot No. 31028 1988–89. –/74 1TD 1W (* –/72, c & t –/70, u –/68, † –/62 1TD 2W). 38.6 t.
DMS. Lot No. 31029 1987–89. –/76 (d –/78, † –/74, t & u –/72). 36.1 t.

156 401	c*	**EM**	P	*EM*	DY	52401 57401
156 402	†cr	**GA**	P	*GA*	NC	52402 57402
156 403	c*	**EM**	P	*EM*	DY	52403 57403
156 404	c*	**EM**	P	*EM*	DY	52404 57404
156 405	c*	**EM**	P	*EM*	DY	52405 57405
156 406	c*	**EM**	P	*EM*	DY	52406 57406
156 407	†cr	**GA**	P	*GA*	NC	52407 57407
156 408	c*	**EM**	P	*EM*	DY	52408 57408
156 409	†cr	**GA**	P	*GA*	NC	52409 57409
156 410	c*	**EM**	P	*EM*	DY	52410 57410
156 411	c*	**EM**	P	*EM*	DY	52411 57411
156 412	†cr	**GA**	P	*GA*	NC	52412 57412
156 413	c*	**EM**	P	*EM*	DY	52413 57413
156 414	c*	**EM**	P	*EM*	DY	52414 57414
156 415	c*	**EM**	P	*EM*	DY	52415 57415
156 416	†cr	**GA**	P	*GA*	NC	52416 57416
156 417	†cr	**GA**	P	*GA*	NC	52417 57417
156 418	†cr	**GA**	P	*GA*	NC	52418 57418
156 419	†cr	**GA**	P	*GA*	NC	52419 57419
156 420	c	**NO**	P	*NO*	AN	52420 57420
156 421	c	**NO**	P	*NO*	AN	52421 57421
156 422	†cr	**GA**	P	*GA*	NC	52422 57422
156 423	c	**NO**	P	*NO*	AN	52423 57423
156 424	c	**NO**	P	*NO*	AN	52424 57424
156 425	c	**NO**	P	*NO*	AN	52425 57425
156 426	c	**NO**	P	*NO*	AN	52426 57426
156 427	c	**NO**	P	*NO*	AN	52427 57427
156 428	c	**NO**	P	*NO*	AN	52428 57428
156 429	c	**NO**	P	*NO*	AN	52429 57429
156 430	t	**SR**	A	*SR*	CK	52430 57430
156 431	t	**SR**	A	*SR*	CK	52431 57431
156 432	t	**SR**	A	*SR*	CK	52432 57432
156 433	t	**SR**	A	*SR*	CK	52433 57433
156 434	t	**SR**	A	*SR*	CK	52434 57434
156 435	t	**SR**	A	*SR*	CK	52435 57435
156 436	†	**SR**	A	*SR*	CK	52436 57436
156 437	t	**SR**	A	*SR*	CK	52437 57437

156 438	d	**NO**	A	*NO*	HT	52438	57438
156 439	t	**SR**	A	*SR*	CK	52439	57439
156 440	c	**NO**	P	*NO*	AN	52440	57440
156 441	c	**NO**	P	*NO*	AN	52441	57441
156 442	t	**SR**	A	*SR*	CK	52442	57442
156 443	d	**NO**	A	*NO*	HT	52443	57443
156 444	d	**NO**	A	*NO*	HT	52444	57444
156 445	ru	**SR**	A	*SR*	CK	52445	57445
156 446	t	**FS**	A	*SR*	CK	52446	57446
156 447	ru	**FS**	A	*SR*	CK	52447	57447
156 448	d	**NO**	A	*NO*	HT	52448	57448
156 449	u	**FS**	A	*SR*	CK	52449	57449
156 450	ru	**FS**	A	*SR*	CK	52450	57450
156 451	d	**NO**	A	*NO*	HT	52451	57451
156 452	c	**NO**	P	*NO*	AN	52452	57452
156 453	ru	**FS**	A	*SR*	CK	52453	57453
156 454	d	**NO**	A	*NO*	HT	52454	57454
156 455	c	**NO**	P	*NO*	AN	52455	57455
156 456	rt	**FS**	A	*SR*	CK	52456	57456
156 457	rt	**FS**	A	*SR*	CK	52457	57457
156 458	rt	**FS**	A	*SR*	CK	52458	57458
156 459	c	**NO**	P	*NO*	AN	52459	57459
156 460	c	**NO**	P	*NO*	AN	52460	57460
156 461	c	**NO**	P	*NO*	AN	52461	57461
156 462		**FS**	A	*SR*	CK	52462	57462
156 463	d	**NO**	A	*NO*	HT	52463	57463
156 464	c	**NO**	P	*NO*	AN	52464	57464
156 465	ru	**FS**	A	*SR*	CK	52465	57465
156 466	c	**NO**	P	*NO*	AN	52466	57466
156 467	r	**FS**	A	*SR*	CK	52467	57467
156 468	d	**NO**	A	*NO*	AN	52468	57468
156 469	d	**NO**	A	*NO*	HT	52469	57469
156 470	c	**EM**	A	*EM*	DY	52470	57470
156 471	d	**NO**	A	*NO*	AN	52471	57471
156 472	d	**NO**	A	*NO*	AN	52472	57472
156 473	c	**EM**	A	*EM*	DY	52473	57473
156 474	rt	**FS**	A	*SR*	CK	52474	57474
156 475	d	**NO**	A	*NO*	HT	52475	57475
156 476	rt	**FS**	A	*SR*	CK	52476	57476
156 477	t	**FS**	A	*SR*	CK	52477	57477
156 478	rt	**FS**	A	*SR*	CK	52478	57478
156 479	d	**NO**	A	*NO*	HT	52479	57479
156 480	d	**NO**	A	*NO*	HT	52480	57480
156 481	d	**NO**	A	*NO*	HT	52481	57481
156 482	d	**NO**	A	*NO*	AN	52482	57482
156 483	d	**NO**	A	*NO*	AN	52483	57483
156 484	d	**NO**	A	*NO*	HT	52484	57484
156 485	ru	**FS**	A	*SR*	CK	52485	57485
156 486	d	**NO**	A	*NO*	AN	52486	57486
156 487	d	**NO**	A	*NO*	AN	52487	57487
156 488	d	**NO**	A	*NO*	AN	52488	57488

156 489	d	**NO**	A	*NO*	AN	52489	57489
156 490	d	**NO**	A	*NO*	HT	52490	57490
156 491	d	**NO**	A	*NO*	AN	52491	57491
156 492	r*	**SR**	A	*SR*	CK	52492	57492
156 493	rt	**FS**	A	*SR*	CK	52493	57493
156 494	u	**SR**	A	*SR*	CK	52494	57494
156 495	u	**SR**	A	*SR*	CK	52495	57495
156 496	u	**FS**	A	*SR*	CK	52496	57496
156 497	c	**EM**	A	*EM*	DY	52497	57497
156 498	c	**EM**	A	*EM*	DY	52498	57498
156 499	rt	**SR**	A	*SR*	CK	52499	57499
156 500	ru	**SR**	A	*SR*	CK	52500	57500
156 501		**SR**	A	*SR*	CK	52501	57501
156 502		**SR**	A	*SR*	CK	52502	57502
156 503		**SR**	A	*SR*	CK	52503	57503
156 504		**SR**	A	*SR*	CK	52504	57504
156 505		**SR**	A	*SR*	CK	52505	57505
156 506		**SR**	A	*SR*	CK	52506	57506
156 507		**SR**	A	*SR*	CK	52507	57507
156 508		**SR**	A	*SR*	CK	52508	57508
156 509		**SR**	A	*SR*	CK	52509	57509
156 510		**SR**	A	*SR*	CK	52510	57510
156 511		**SR**	A	*SR*	CK	52511	57511
156 512		**SR**	A	*SR*	CK	52512	57512
156 513		**SR**	A	*SR*	CK	52513	57513
156 514		**SR**	A	*SR*	CK	52514	57514

Names:

156 416	Saint Edmund
156 420	LA' AL RATTY Ravenglass & Eskdale Railway
156 438	Timothy Hackworth
156 440	George Bradshaw
156 441	William Huskisson MP
156 444	Councillor Bill Cameron
156 459	Benny Rothman – The Manchester Rambler
156 460	Driver John Axon G.C.
156 464	Lancashire DalesRail
156 466	Gracie Fields
156 482	Elizabeth Gaskell
156 490	Captain James Cook Master Mariner

CLASS 158/0   BREL

DMSL(B)–DMSL(A) or DMCL–DMSL or DMSL–MSL–DMSL.

Construction: Welded aluminium.
Engines: 158 701–813/158 880–890/158 950–961: One Cummins NTA855R1
of 260 kW (350 hp) at 2100 rpm.
158 815–862: One Perkins 2006-TWH of 260 kW (350 hp) at 2100 rpm.
158 863–872: One Cummins NTA855R3 of 300 kW (400 hp) at 1900 rpm.
Bogies: One BREL P4 (powered) and one BREL T4 (non-powered) per car.
Couplers: BSI.   **Dimensions:** 22.57 x 2.70 m.
Gangways: Throughout.   **Wheel Arrangement:** 2-B + B-2.
Doors: Twin-leaf swing plug.   **Maximum Speed:** 90 mph.
Seating Layout: 2+2 facing/unidirectional in all Standard and First Class
except 2+1 in South West Trains First Class.
Multiple Working: Within class and with Classes 142, 143, 144, 150, 153,
155, 156, 159, 170 and 172.

ScotRail 158s 158 701–741 are "fitted" for RETB. When a unit arrives at
Inverness the cab display unit is clipped on and plugged in.

Arriva Trains Wales units have ERTMS plugged in at Machynlleth for
working the Cambrian Lines.

*   Refurbished ScotRail units fitted with Grammer seating, additional
    luggage racks and cycle stowage areas.
    ScotRail units 158 726–741 are fitted with Richmond seating.
†   Refurbished East Midlands Trains units with Grammer seating.
c   Chapman seating.
s   Refurbished Arriva Trains Wales units with Grammer seating.
u   Refurbished former South West Trains units with Class 159-style
    interiors, including First Class seating.
w   Refurbished First Great Western units. Units 158 745–749/751/762/767
    (now formed into 3-car sets) have been fitted with Richmond seating.

Northern promotional vinyls:

158 784 PTEG: 40 years.
158 787, 158 792–796 Sheffield–Leeds fast service.
158 790 Rugby League (Northern Rail Cup).
158 860 Keighley & Brontë Country.
158 901–910 Leeds–Bradford–Manchester route (the "Calder Valley").

DMSL(B). Lot No. 31051 BREL Derby 1989–92. –/68 1TD 1W. († –/72 1TD 1W,
c, w –/66 1TD 1W, s –/64(4) 1TD 2W, t –/64 1TD 1W). 38.5 t.
MSL. Lot No. 31050 BREL Derby 1991. –/66(3) 1T. 38.5 t.
DMSL(A). Lot No. 31052 BREL Derby 1989–92. –/70 1T († –/74, c, w –/68 1T,
* –/64(2) 1T plus cycle stowage area, s –/70 1T, t –/66 1T). 38.5 t.

The above details refer to the "as built" condition. The following DMSL(B)
have now been converted to DMCL as follows:

52701–736/52738–741 (ScotRail). 15/53 1TD 1W (* refurbished sets 14/46(6)
1TD 1W plus cycle stowage area).
52786/789 (Former South West Trains units). 13/44 1TD 1W.

158 701	*	**FS**	P	*SR*	IS	52701	57701
158 702	*	**FS**	P	*SR*	IS	52702	57702
158 703	*	**FS**	P	*SR*	IS	52703	57703
158 704	*	**FS**	P	*SR*	IS	52704	57704
158 705	*	**FS**	P	*SR*	IS	52705	57705
158 706	*	**FS**	P	*SR*	IS	52706	57706
158 707	*	**FS**	P	*SR*	IS	52707	57707
158 708	*	**FS**	P	*SR*	IS	52708	57708
158 709	*	**FS**	P	*SR*	IS	52709	57709
158 710	*	**FS**	P	*SR*	IS	52710	57710
158 711	*	**FS**	P	*SR*	IS	52711	57711
158 712	*	**FS**	P	*SR*	IS	52712	57712
158 713	*	**FS**	P	*SR*	IS	52713	57713
158 714	*	**FS**	P	*SR*	IS	52714	57714
158 715	*	**FS**	P	*SR*	IS	52715	57715
158 716	*	**FS**	P	*SR*	IS	52716	57716
158 717	*	**FS**	P	*SR*	IS	52717	57717
158 718	*	**FS**	P	*SR*	IS	52718	57718
158 719	*	**FS**	P	*SR*	IS	52719	57719
158 720	*	**FS**	P	*SR*	IS	52720	57720
158 721	*	**FS**	P	*SR*	IS	52721	57721
158 722	*	**FS**	P	*SR*	IS	52722	57722
158 723	*	**FS**	P	*SR*	IS	52723	57723
158 724	*	**FS**	P	*SR*	IS	52724	57724
158 725	*	**FS**	P	*SR*	IS	52725	57725
158 726		**FS**	P	*SR*	HA	52726	57726
158 727		**FS**	P	*SR*	HA	52727	57727
158 728		**FS**	P	*SR*	HA	52728	57728
158 729		**FS**	P	*SR*	HA	52729	57729
158 730		**FS**	P	*SR*	HA	52730	57730
158 731		**FS**	P	*SR*	HA	52731	57731
158 732		**FS**	P	*SR*	HA	52732	57732
158 733		**FS**	P	*SR*	HA	52733	57733
158 734		**FS**	P	*SR*	HA	52734	57734
158 735		**FS**	P	*SR*	HA	52735	57735
158 736		**FS**	P	*SR*	HA	52736	57736
158 738		**FS**	P	*SR*	HA	52738	57738
158 739		**FS**	P	*SR*	HA	52739	57739
158 740		**FS**	P	*SR*	HA	52740	57740
158 741		**FS**	P	*SR*	HA	52741	57741
158 752		**NO**	P	*NO*	NL	52752	58716	57752
158 753		**NO**	P	*NO*	NL	52753	58710	57753
158 754		**NO**	P	*NO*	NL	52754	58708	57754
158 755		**NO**	P	*NO*	NL	52755	58702	57755
158 756		**NO**	P	*NO*	NL	52756	58712	57756
158 757		**NO**	P	*NO*	NL	52757	58706	57757
158 758		**NO**	P	*NO*	NL	52758	58714	57758
158 759		**NO**	P	*NO*	NL	52759	58713	57759
158 763	w	**FI**	P	*GW*	PM	52763	57763
158 766	w	**FI**	P	*GW*	PM	52766	57766
158 770	†	**ST**	P	*EM*	NM	52770	57770

158 773	†	**ST**	P	*EM*	NM	52773	57773	
158 774	†	**ST**	P	*EM*	NM	52774	57774	
158 777	†	**ST**	P	*EM*	NM	52777	57777	
158 780	†	**ST**	A	*EM*	NM	52780	57780	
158 782		**SR**	A	*SR*	HA	52782	57782	
158 783	†	**ST**	A	*EM*	NM	52783	57783	
158 784		**NO**	A	*NO*	NL	52784	57784	
158 785	†	**ST**	A	*EM*	NM	52785	57785	
158 786	u	**SR**	A	*SR*	HA	52786	57786	
158 787		**NO**	A	*NO*	NL	52787	57787	
158 788	†	**ST**	A	*EM*	NM	52788	57788	
158 789	u	**SR**	A	*SR*	HA	52789	57789	
158 790		**NO**	A	*NO*	NL	52790	57790	
158 791		**NO**	A	*NO*	NL	52791	57791	
158 792		**NO**	A	*NO*	NL	52792	57792	
158 793		**NO**	A	*NO*	NL	52793	57793	
158 794		**NO**	A	*NO*	NL	52794	57794	
158 795		**NO**	A	*NO*	NL	52795	57795	
158 796		**NO**	A	*NO*	NL	52796	57796	
158 797		**NO**	A	*NO*	NL	52797	57797	
158 798	w	**FI**	P	*GW*	PM	52798	58715	57798
158 799	†	**ST**	P	*EM*	NM	52799	57799	
158 806	†	**ST**	P	*EM*	NM	52806	57806	
158 810	†	**ST**	P	*EM*	NM	52810	57810	
158 812	†	**ST**	P	*EM*	NM	52812	57812	
158 813	†	**ST**	P	*EM*	NM	52813	57813	
158 815	c	**NO**	A	*NO*	NL	52815	57815	
158 816	c	**NO**	A	*NO*	NL	52816	57816	
158 817	c	**NO**	A	*NO*	NL	52817	57817	
158 818	es	**AW**	A	*AW*	MN	52818	57818	
158 819	es	**AW**	A	*AW*	MN	52819	57819	
158 820	es	**AW**	A	*AW*	MN	52820	57820	
158 821	es	**AW**	A	*AW*	MN	52821	57821	
158 822	es	**AW**	A	*AW*	MN	52822	57822	
158 823	es	**AW**	A	*AW*	MN	52823	57823	
158 824	es	**AW**	A	*AW*	MN	52824	57824	
158 825	es	**AW**	A	*AW*	MN	52825	57825	
158 826	es	**AW**	A	*AW*	MN	52826	57826	
158 827	es	**AW**	A	*AW*	MN	52827	57827	
158 828	es	**AW**	A	*AW*	MN	52828	57828	
158 829	es	**AW**	A	*AW*	MN	52829	57829	
158 830	es	**AW**	A	*AW*	MN	52830	57830	
158 831	es	**AW**	A	*AW*	MN	52831	57831	
158 832	es	**AW**	A	*AW*	MN	52832	57832	
158 833	es	**AW**	A	*AW*	MN	52833	57833	
158 834	es	**AW**	A	*AW*	MN	52834	57834	
158 835	es	**AW**	A	*AW*	MN	52835	57835	
158 836	es	**AW**	A	*AW*	MN	52836	57836	
158 837	es	**AW**	A	*AW*	MN	52837	57837	
158 838	es	**AW**	A	*AW*	MN	52838	57838	
158 839	es	**AW**	A	*AW*	MN	52839	57839	

158 840	es	**AW**	A	*AW*	MN	52840	57840
158 841	es	**AW**	A	*AW*	MN	52841	57841
158 842	c	**NO**	A	*NO*	NL	52842	57842
158 843	c	**NO**	A	*NO*	NL	52843	57843
158 844		**NO**	A	*NO*	NL	52844	57844
158 845		**NO**	A	*NO*	NL	52845	57845
158 846	†	**ST**	A	*EM*	NM	52846	57846
158 847	†	**ST**	A	*EM*	NM	52847	57847
158 848		**NO**	A	*NO*	NL	52848	57848
158 849		**NO**	A	*NO*	NL	52849	57849
158 850		**NO**	A	*NO*	NL	52850	57850
158 851		**NO**	A	*NO*	NL	52851	57851
158 852	†	**ST**	A	*EM*	NM	52852	57852
158 853		**NO**	A	*NO*	NL	52853	57853
158 854	†	**ST**	A	*EM*	NM	52854	57854
158 855		**NO**	A	*NO*	NL	52855	57855
158 856	†	**ST**	A	*EM*	NM	52856	57856
158 857	†	**ST**	A	*EM*	NM	52857	57857
158 858	†	**ST**	A	*EM*	NM	52858	57858
158 859		**NO**	A	*NO*	NL	52859	57859
158 860		**NO**	A	*NO*	NL	52860	57860
158 861		**NO**	A	*NO*	NL	52861	57861
158 862	†	**ST**	A	*EM*	NM	52862	57862
158 863	†	**ST**	A	*EM*	NM	52863	57863
158 864	†	**ST**	A	*EM*	NM	52864	57864
158 865	†	**ST**	A	*EM*	NM	52865	57865
158 866	†	**ST**	A	*EM*	NM	52866	57866
158 867	c	**SR**	A	*SR*	HA	52867	57867
158 868	c	**SR**	A	*SR*	HA	52868	57868
158 869	c	**SR**	A	*SR*	HA	52869	57869
158 870	c	**SR**	A	*SR*	HA	52870	57870
158 871	c	**SR**	A	*SR*	HA	52871	57871
158 872	c	**NO**	A	*NO*	NL	52872	57872

Names (ScotRail units carry their names on the unit ends):

158 702	BBC Scotland 75 years
158 707	Far North Line 125th ANNIVERSARY
158 715	Haymarket
158 720	Inverness & Nairn Railway – 150 years
158 784	Barbara Castle
158 791	County of Nottinghamshire
158 796	Fred Trueman Cricketing Legend
158 797	Jane Tomlinson
158 860	Ian Dewhirst

Class 158/8. Refurbished South West Trains units. Converted from former TransPennine Express units at Wabtec, Doncaster in 2007. 2+1 seating in First Class. Details as Class 158/0 except:

DMCL. Lot No. 31051 BREL Derby 1989–92. 13/44 1TD 1W. 38.5 t.
DMSL. Lot No. 31052 BREL Derby 1989–92. –/70 1T. 38.5 t.

158 880	(158 737)	**ST**	P	*SW*	SA	52737	57737
158 881	(158 742)	**ST**	P	*SW*	SA	52742	57742
158 882	(158 743)	**ST**	P	*SW*	SA	52743	57743
158 883	(158 744)	**ST**	P	*SW*	SA	52744	57744
158 884	(158 772)	**ST**	P	*SW*	SA	52772	57772
158 885	(158 775)	**ST**	P	*SW*	SA	52775	57775
158 886	(158 779)	**ST**	P	*SW*	SA	52779	57779
158 887	(158 781)	**ST**	P	*SW*	SA	52781	57781
158 888	(158 802)	**ST**	P	*SW*	SA	52802	57802
158 889	(158 808)	**ST**	P	*SW*	SA	52808	57808
158 890	(158 814)	**ST**	P	*SW*	SA	52814	57814

CLASS 158/9   BREL

DMSL–DMS. Units leased by West Yorkshire PTE but managed by Eversholt Rail. Details as Class 158/0 except for seating and toilets.

DMSL. Lot No. 31051 BREL Derby 1990–92. –/70 1TD 1W. 38.5 t.
DMS. Lot No. 31052 BREL Derby 1990–92. –/72 and parcels area. 38.5 t.

158 901	**NO**	E	*NO*	NL	52901	57901	
158 902	**NO**	E	*NO*	NL	52902	57902	
158 903	**NO**	E	*NO*	NL	52903	57903	
158 904	**NO**	E	*NO*	NL	52904	57904	
158 905	**NO**	E	*NO*	NL	52905	57905	
158 906	**NO**	E	*NO*	NL	52906	57906	
158 907	**NO**	E	*NO*	NL	52907	57907	
158 908	**NO**	E	*NO*	NL	52908	57908	
158 909	**NO**	E	*NO*	NL	52909	57909	
158 910	**NO**	E	*NO*	NL	52910	57910	William Wilberforce

CLASS 158/0   BREL

DMSL–DMSL–DMSL. Refurbished units reformed for First Great Western. For vehicle details see above. Formations can be flexible depending on when unit exams become due.

158 950	w	**FI**	P	*GW*	PM	57751	52761	57761
158 951	w	**FI**	P	*GW*	PM	52751	52764	57764
158 952	w	**FI**	P	*GW*	PM	57745	52762	57762
158 953	w	**FI**	P	*GW*	PM	52745	52750	57750
158 954	w	**FI**	P	*GW*	PM	57747	52760	57760
158 955	w	**FI**	P	*GW*	PM	52747	52765	57765
158 956	w	**FI**	P	*GW*	PM	52748	52768	57768
158 957	w	**FI**	P	*GW*	PM	57748	52771	57771
158 958	w	**FI**	P	*GW*	PM	57746	52776	57776
158 959	w	**FI**	P	*GW*	PM	52746	52778	57778
158 960	w	**FI**	P	*GW*	PM	57749	52769	57769
158 961	w	**FI**	P	*GW*	PM	52749	52767	57767

CLASS 159/0   BREL

DMCL–MSL–DMSL. Built as Class 158. Converted before entering passenger service to Class 159 by Rosyth Dockyard.

Construction: Welded aluminium.
Engines: One Cummins NTA855R3 of 300 kW (400 hp) at 1900 rpm.
Bogies: One BREL P4 (powered) and one BREL T4 (non-powered) per car.
Couplers: BSI.   **Dimensions:** 22.16 x 2.70 m.
Gangways: Throughout.   **Wheel Arrangement:** 2-B + B-2 + B-2.
Doors: Twin-leaf swing plug.   **Maximum Speed:** 90 mph.
Seating Layout: 1: 2+1 facing, 2: 2+2 facing/unidirectional.
Multiple Working: Within class and with Classes 142, 143, 144, 150, 153, 155, 156, 158 and 170.

DMCL. Lot No. 31051 BREL Derby 1992–93. 23/28 1TD 1W. 38.5 t.
MSL. Lot No. 31050 BREL Derby 1992–93. –/70(6) 1T. 38.5 t.
DMSL. Lot No. 31052 BREL Derby 1992–93. –/72 1T. 38.5 t.

159 001	**ST**	P	*SW*	SA	52873	58718	57873	CITY OF EXETER
159 002	**ST**	P	*SW*	SA	52874	58719	57874	CITY OF SALISBURY
159 003	**ST**	P	*SW*	SA	52875	58720	57875	TEMPLECOMBE
159 004	**ST**	P	*SW*	SA	52876	58721	57876	BASINGSTOKE AND DEANE
159 005	**ST**	P	*SW*	SA	52877	58722	57877	WEST OF ENGLAND LINE
159 006	**ST**	P	*SW*	SA	52878	58723	57878	THE SEATON TRAMWAY Seaton–Colyford–Colyton
159 007	**ST**	P	*SW*	SA	52879	58724	57879	
159 008	**ST**	P	*SW*	SA	52880	58725	57880	
159 009	**ST**	P	*SW*	SA	52881	58726	57881	
159 010	**ST**	P	*SW*	SA	52882	58727	57882	
159 011	**ST**	P	*SW*	SA	52883	58728	57883	
159 012	**ST**	P	*SW*	SA	52884	58729	57884	
159 013	**ST**	P	*SW*	SA	52885	58730	57885	
159 014	**ST**	P	*SW*	SA	52886	58731	57886	
159 015	**ST**	P	*SW*	SA	52887	58732	57887	
159 016	**ST**	P	*SW*	SA	52888	58733	57888	
159 017	**ST**	P	*SW*	SA	52889	58734	57889	
159 018	**ST**	P	*SW*	SA	52890	58735	57890	
159 019	**ST**	P	*SW*	SA	52891	58736	57891	
159 020	**ST**	P	*SW*	SA	52892	58737	57892	
159 021	**ST**	P	*SW*	SA	52893	58738	57893	
159 022	**ST**	P	*SW*	SA	52894	58739	57894	

CLASS 159/1   BREL

DMCL–MSL–DMSL. Units converted from Class 158s at Wabtec, Doncaster in 2006–07 for South West Trains.

Details as Class 158/0 except:
Seating Layout: 1: 2+1 facing, 2: 2+2 facing/unidirectional.

DMCL. Lot No. 31051 BREL Derby 1989–92. 24/28 1TD 1W. 38.5 t.
MSL. Lot No. 31050 BREL Derby 1989–92. –/70 1T. 38.5 t.
DMSL. Lot No. 31052 BREL Derby 1989–92. –/72 1T.38.5 t.

159 101	(158 800)	**ST**	P	*SW*	SA	52800	58717	57800
159 102	(158 803)	**ST**	P	*SW*	SA	52803	58703	57803
159 103	(158 804)	**ST**	P	*SW*	SA	52804	58704	57804
159 104	(158 805)	**ST**	P	*SW*	SA	52805	58705	57805
159 105	(158 807)	**ST**	P	*SW*	SA	52807	58707	57807
159 106	(158 809)	**ST**	P	*SW*	SA	52809	58709	57809
159 107	(158 811)	**ST**	P	*SW*	SA	52811	58711	57811
159 108	(158 801)	**ST**	P	*SW*	SA	52801	58701	57801

CLASS 165/0   NETWORK TURBO   BREL

DMSL–DMS and DMSL–MS–DMS. Chiltern Railways units. Refurbished 2003–05 with First Class seats removed and air conditioning fitted.
Construction: Welded aluminium.
Engines: One Perkins 2006-TWH of 260 kW (350 hp) at 2100 rpm.
Bogies: BREL P3-17 (powered), BREL T3-17 (non-powered).
Couplers: BSI.
Dimensions: 23.50/23.25 x 2.81 m.
Gangways: Within unit only.   **Wheel Arrangement:** 2-B (+ B-2) + B-2.
Doors: Twin-leaf swing plug.   **Maximum Speed:** 75 mph.
Seating Layout: 2+2/3+2 facing/unidirectional.
Multiple Working: Within class and with Classes 166, 168, 170 and 172.

Fitted with tripcocks for working over London Underground tracks between Harrow-on-the-Hill and Amersham.

58801–822/58873–878. DMSL. Lot No. 31087 BREL York 1990. –/82(7) 1T 2W. 40.1 t.
58823–833. DMSL. Lot No. 31089 BREL York 1991–92. –/82(7) 1T 2W. 40.1 t.
MS. Lot No. 31090 BREL York 1991–92. –/106. 37.0 t.
DMS. Lot No. 31088 BREL York 1991–92. –/94. 39.4 t.

165 001	**CR**	A	*CR*	AL	58801	58834
165 002	**CR**	A	*CR*	AL	58802	58835
165 003	**CR**	A	*CR*	AL	58803	58836
165 004	**CR**	A	*CR*	AL	58804	58837
165 005	**CR**	A	*CR*	AL	58805	58838
165 006	**CR**	A	*CR*	AL	58806	58839
165 007	**CR**	A	*CR*	AL	58807	58840
165 008	**CR**	A	*CR*	AL	58808	58841
165 009	**CR**	A	*CR*	AL	58809	58842
165 010	**CR**	A	*CR*	AL	58810	58843
165 011	**CR**	A	*CR*	AL	58811	58844
165 012	**CR**	A	*CR*	AL	58812	58845
165 013	**CR**	A	*CR*	AL	58813	58846
165 014	**CR**	A	*CR*	AL	58814	58847
165 015	**CR**	A	*CR*	AL	58815	58848
165 016	**CR**	A	*CR*	AL	58816	58849
165 017	**CR**	A	*CR*	AL	58817	58850

165 018	**CR**	A	*CR*	AL	58818		58851
165 019	**CR**	A	*CR*	AL	58819		58852
165 020	**CR**	A	*CR*	AL	58820		58853
165 021	**CR**	A	*CR*	AL	58821		58854
165 022	**CR**	A	*CR*	AL	58822		58855
165 023	**CR**	A	*CR*	AL	58873		58867
165 024	**CR**	A	*CR*	AL	58874		58868
165 025	**CR**	A	*CR*	AL	58875		58869
165 026	**CR**	A	*CR*	AL	58876		58870
165 027	**CR**	A	*CR*	AL	58877		58871
165 028	**CR**	A	*CR*	AL	58878		58872
165 029	**CR**	A	*CR*	AL	58823	55404	58856
165 030	**CR**	A	*CR*	AL	58824	55405	58857
165 031	**CR**	A	*CR*	AL	58825	55406	58858
165 032	**CR**	A	*CR*	AL	58826	55407	58859
165 033	**CR**	A	*CR*	AL	58827	55408	58860
165 034	**CR**	A	*CR*	AL	58828	55409	58861
165 035	**CR**	A	*CR*	AL	58829	55410	58862
165 036	**CR**	A	*CR*	AL	58830	55411	58863
165 037	**CR**	A	*CR*	AL	58831	55412	58864
165 038	**CR**	A	*CR*	AL	58832	55413	58865
165 039	**CR**	A	*CR*	AL	58833	55414	58866

CLASS 165/1 NETWORK TURBO BREL

First Great Western units. DMCL–MS–DMS or DMCL–DMS.

Construction: Welded aluminium.
Engines: One Perkins 2006-TWH of 260 kW (350 hp) at 2100 rpm.
Bogies: BREL P3-17 (powered), BREL T3-17 (non-powered).
Couplers: BSI.
Dimensions: 23.50/23.25 x 2.81 m.
Gangways: Within unit only.   **Wheel Arrangement:** 2-B (+ B-2) + B-2.
Doors: Twin-leaf swing plug.   **Maximum Speed:** 90 mph.
Seating Layout: 1: 2+2 facing, 2: 3+2 facing/unidirectional.
Multiple Working: Within class and with Classes 166, 168, 170 and 172.

58953–969. DMCL. Lot No. 31098 BREL York 1992. 16/66 1T. 38.0 t.
58879–898. DMCL. Lot No. 31096 BREL York 1992. 16/72 1T. 38.0 t.
MS. Lot No. 31099 BREL 1992. –/106. 37.0 t.
DMS. Lot No. 31097 BREL 1992. –/98. 37.0 t.

165 101	**FD**	A	*GW*	RG	58953	55415	58916
165 102	**FD**	A	*GW*	RG	58954	55416	58917
165 103	**FD**	A	*GW*	RG	58955	55417	58918
165 104	**FD**	A	*GW*	RG	58956	55418	58919
165 105	**FD**	A	*GW*	RG	58957	55419	58920
165 106	**FD**	A	*GW*	RG	58958	55420	58921
165 107	**FD**	A	*GW*	RG	58959	55421	58922
165 108	**FD**	A	*GW*	RG	58960	55422	58923
165 109	**FD**	A	*GW*	RG	58961	55423	58924
165 110	**FD**	A	*GW*	RG	58962	55424	58925

165 111	**FD**	A	*GW*	RG	58963	55425	58926
165 112	**FD**	A	*GW*	RG	58964	55426	58927
165 113	**FD**	A	*GW*	RG	58965	55427	58928
165 114	**FD**	A	*GW*	RG	58966	55428	58929
165 116	**FD**	A	*GW*	RG	58968	55430	58931
165 117	**FD**	A	*GW*	RG	58969	55431	58932
165 118	**FD**	A	*GW*	RG	58879		58933
165 119	**FD**	A	*GW*	RG	58880		58934
165 120	**FD**	A	*GW*	RG	58881		58935
165 121	**FD**	A	*GW*	RG	58882		58936
165 122	**FD**	A	*GW*	RG	58883		58937
165 123	**FD**	A	*GW*	RG	58884		58938
165 124	**FD**	A	*GW*	RG	58885		58939
165 125	**FD**	A	*GW*	RG	58886		58940
165 126	**FD**	A	*GW*	RG	58887		58941
165 127	**FD**	A	*GW*	RG	58888		58942
165 128	**FD**	A	*GW*	RG	58889		58943
165 129	**FD**	A	*GW*	RG	58890		58944
165 130	**FD**	A	*GW*	RG	58891		58945
165 131	**FD**	A	*GW*	RG	58892		58946
165 132	**FD**	A	*GW*	RG	58893		58947
165 133	**FD**	A	*GW*	RG	58894		58948
165 134	**FD**	A	*GW*	RG	58895		58949
165 135	**FD**	A	*GW*	RG	58896		58950
165 136	**FD**	A	*GW*	RG	58897		58951
165 137	**FD**	A	*GW*	RG	58898		58952

CLASS 166   NETWORK EXPRESS TURBO   ABB

DMSL–MS–DMCL. First Great Western units, built for Paddington–Oxford/
Newbury services. Air conditioned and with additional luggage space
compared to the Class 165s. The DMSL vehicles have had their 16 First
Class seats declassified.

Construction: Welded aluminium.
Engines: One Perkins 2006-TWH of 260 kW (350 hp) at 2100 rpm.
Bogies: BREL P3-17 (powered), BREL T3-17 (non-powered).
Couplers: BSI.
Dimensions: 23.50 x 2.81 m.

Gangways: Within unit only.	**Wheel Arrangement:** 2-B + B-2 + B-2.
Doors: Twin-leaf swing plug.	**Maximum Speed:** 90 mph.

Seating Layout: 1: 2+2 facing, 2: 2+2/3+2 facing/unidirectional.
Multiple Working: Within class and with Classes 165, 168, 170 and 172.

* Refurbished with a new universal access toilet to comply with the 2020
accessibility regulations. Full details awaited.

DMSL. Lot No. 31116 ABB York 1992–93. –/84 1T. 39.6 t.
MS. Lot No. 31117 ABB York 1992–93. –/91. 38.0 t.
DMCL. Lot No. 31116 ABB York 1992–93. 16/68 1T. 39.6 t.

166 201	**FD**	A	*GW*	RG	58101	58601	58122
166 202	**FD**	A	*GW*	RG	58102	58602	58123

166 203	**FD**	A	*GW*	RG	58103	58603	58124
166 204	**FD**	A	*GW*	RG	58104	58604	58125
166 205	**FD**	A	*GW*	RG	58105	58605	58126
166 206	**FD**	A	*GW*	RG	58106	58606	58127
166 207	**FD**	A	*GW*	RG	58107	58607	58128
166 208	**FD**	A	*GW*	RG	58108	58608	58129
166 209	**FD**	A	*GW*	RG	58109	58609	58130
166 210	**FD**	A	*GW*	RG	58110	58610	58131
166 211	**FD**	A	*GW*	RG	58111	58611	58132
166 212	**FD**	A	*GW*	RG	58112	58612	58133
166 213	**FD**	A	*GW*	RG	58113	58613	58134
166 214	**FD**	A	*GW*	RG	58114	58614	58135
166 215	**FD**	A	*GW*	RG	58115	58615	58136
166 216	**FD**	A	*GW*	RG	58116	58616	58137
166 217	**FD**	A	*GW*	RG	58117	58617	58138
166 218	**FD**	A	*GW*	RG	58118	58618	58139
166 219	**FD**	A	*GW*	RG	58119	58619	58140
166 220	**FD**	A	*GW*	RG	58120	58620	58141
166 221	* **FB**	A	*GW*	RG	58121	58621	58142

Name: 166 221 Reading Train Care Depot/READING TRAIN CARE DEPOT *(alt sides)*

CLASS 168   CLUBMAN   ADTRANZ/BOMBARDIER

Air conditioned.

Construction: Welded aluminium bodies with bolt-on steel ends.
Engines: One MTU 6R183TD13H of 315 kW (422 hp) at 1900 rpm.
Transmission: Hydraulic. Voith T211rzze to ZF final drive.
Bogies: One Adtranz P3–23 and one BREL T3–23 per car.
Couplers: BSI at outer ends, bar within unit.
Dimensions: Class 168/0: 24.10/23.61 x 2.69 m. Others: 23.62/23.61 x 2.69 m.
Gangways: Within unit only.   **Wheel Arrangement:** 2-B (+ B-2 + B-2) + B-2.
Doors: Twin-leaf swing plug.   **Maximum Speed:** 100 mph.
Seating Layout: 2+2 facing/unidirectional.
Multiple Working: Within class and with Classes 165 and 166.

Fitted with tripcocks for working over London Underground tracks between Harrow-on-the-Hill and Amersham.

Class 168/0. Original Design. DMSL(A)–MS–MSL–DMSL(B) or DMSL(A)–MSL–MS–DMSL(B).

58451–455 were numbered 58656–660 for a time when used in 168 106–110.

58151–155. DMSL(A). Adtranz Derby 1997–98. –/57 1TD 1W. 44.0 t.
58651–655. MSL. Adtranz Derby 1998. –/73 1T. 41.0 t.
58451–455. MS. Adtranz Derby 1998. –/77. 41.0 t.
58251–255. DMSL(B). Adtranz Derby 1998. –/68 1T. 43.6 t.

168 001	**CL**	P	*CR*	AL	58151	58451	58651	58251
168 002	**CL**	P	*CR*	AL	58152	58652	58452	58252
168 003	**CL**	P	*CR*	AL	58153	58453	58653	58253
168 004	**CL**	P	*CR*	AL	58154	58654	58454	58254
168 005	**CL**	P	*CR*	AL	58155	58655	58455	58255

Class 168/1. These units are effectively Class 170s. DMSL(A)–MSL–MS–DMSL(B) or DMSL(A)–MS–DMSL(B).

58461–463 have been renumbered from 58661–663.

58156–163. DMSL(A). Adtranz Derby 2000. –/57 1TD 2W. 45.2 t.
58456–460. MS. Bombardier Derby 2002. –/76. 41.8 t.
58756–757. MSL. Bombardier Derby 2002. –/73 1T. 42.9 t.
58461–463. MS. Adtranz Derby 2000. –/76. 42.4 t.
58256–263. DMSL(B). Adtranz Derby 2000. –/69 1T. 45.2 t.

168 106	CL	P	CR	AL	58156	58456	58756	58256
168 107	CR	P	CR	AL	58157	58757	58457	58257
168 108	CR	P	CR	AL	58158		58458	58258
168 109	CL	P	CR	AL	58159		58459	58259
168 110	CL	P	CR	AL	58160		58460	58260
168 111	CR	E	CR	AL	58161		58461	58261
168 112	CR	E	CR	AL	58162		58462	58262
168 113	CR	E	CR	AL	58163		58463	58263

Class 168/2. These units are effectively Class 170s. DMSL(A)–(MS)–MS–DMSL(B).

58164–169. DMSL(A). Bombardier Derby 2003–04. –/57 1TD 2W. 45.4 t.
58365–367. MS. Bombardier Derby 2006. –/76. 43.3 t.
58464/468/469. MS. Bombardier Derby 2003–04. –/76. 44.0 t.
58465–467. MS. Bombardier Derby 2006. –/76. 43.3 t.
58264–269. DMSL(B). Bombardier Derby 2003–04. –/69 1T. 45.5 t.

168 214	CL	P	CR	AL	58164		58464	58264
168 215	CL	P	CR	AL	58165	58365	58465	58265
168 216	CL	P	CR	AL	58166	58366	58466	58266
168 217	CL	P	CR	AL	58167	58367	58467	58267
168 218	CL	P	CR	AL	58168		58468	58268
168 219	CL	P	CR	AL	58169		58469	58269

CLASS 170   TURBOSTAR ADTRANZ/BOMBARDIER

Various formations. Air conditioned.

Construction: Welded aluminium bodies with bolt-on steel ends.
Engines: One MTU 6R183TD13H of 315 kW (422 hp) at 1900 rpm.
Transmission: Hydraulic. Voith T211rzze to ZF final drive.
Bogies: One Adtranz P3–23 and one BREL T3–23 per car.
Couplers: BSI at outer ends, bar within later build units.
Dimensions: 23.62/23.61 x 2.69 m.
Gangways: Within unit only.   **Wheel Arrangement:** 2-B (+ B-2) + B-2.
Doors: Twin-leaf sliding plug.   **Maximum Speed:** 100 mph.
Seating Layout: 1: 2+1 facing/unidirectional. 2: 2+2 unidirectional/facing.
Multiple Working: Within class and with Classes 150, 153, 155, 156, 158, 159 and 172.

Class 170/1. CrossCountry (former Midland Mainline) units. Lazareni seating. DMSL–MS–DMCL/DMSL–DMCL.

DMSL. Adtranz Derby 1998–99. –/59 1TD 2W. 45.0 t.
MS. Adtranz Derby 2001. –/80. 43.0 t.
DMCL. Adtranz Derby 1998–99. 9/52 1T. 44.8 t

170 101	**XC**	P	*XC*	TS	50101	55101	79101
170 102	**XC**	P	*XC*	TS	50102	55102	79102
170 103	**XC**	P	*XC*	TS	50103	55103	79103
170 104	**XC**	P	*XC*	TS	50104	55104	79104
170 105	**XC**	P	*XC*	TS	50105	55105	79105
170 106	**XC**	P	*XC*	TS	50106	55106	79106
170 107	**XC**	P	*XC*	TS	50107	55107	79107
170 108	**XC**	P	*XC*	TS	50108	55108	79108
170 109	**XC**	P	*XC*	TS	50109	55109	79109
170 110	**XC**	P	*XC*	TS	50110	55110	79110
170 111	**XC**	P	*XC*	TS	50111		79111
170 112	**XC**	P	*XC*	TS	50112		79112
170 113	**XC**	P	*XC*	TS	50113		79113
170 114	**XC**	P	*XC*	TS	50114		79114
170 115	**XC**	P	*XC*	TS	50115		79115
170 116	**XC**	P	*XC*	TS	50116		79116
170 117	**XC**	P	*XC*	TS	50117		79117

Class 170/2. Greater Anglia 3-car units. Chapman seating. DMCL–MSL–DMSL.

Advertising livery: 170 208 Breckland Line (Norwich–Cambridge).

DMCL. Adtranz Derby 1999. 7/39 1TD 2W. 45.0 t.
MSL. Adtranz Derby 1999. –/68 1T. Guard's office. 45.3 t.
DMSL. Adtranz Derby 1999. –/66 1T. 43.4 t.

170 201	r	**1**	P	*GA*	NC	50201	56201	79201
170 202	r	**1**	P	*GA*	NC	50202	56202	79202
170 203	r	**1**	P	*GA*	NC	50203	56203	79203
170 204	r	**1**	P	*GA*	NC	50204	56204	79204
170 205	r	**1**	P	*GA*	NC	50205	56205	79205
170 206	r	**1**	P	*GA*	NC	50206	56206	79206
170 207	r	**1**	P	*GA*	NC	50207	56207	79207
170 208	r	**AL**	P	*GA*	NC	50208	56208	79208

Class 170/2. Greater Anglia 2-car units. Chapman seating. DMSL–DMCL.

DMSL. Bombardier Derby 2002. –/57 1TD 2W. 45.7 t.
DMCL. Bombardier Derby 2002. 9/53 1T. 45.7 t.

170 270	r	**1**	P	*GA*	NC	50270	79270
170 271	r	**AN**	P	*GA*	NC	50271	79271
170 272	r	**AN**	P	*GA*	NC	50272	79272
170 273	r	**AN**	P	*GA*	NC	50273	79273

Class 170/3. TransPennine Express units. Chapman seating. DMCL–DMSL. 170 309 renumbered from 170 399.

50301–308/399. DMCL. Adtranz Derby 2000–01. 8/43 1TD 2W. 45.8 t.
79301–308/399. DMSL. Adtranz Derby 2000–01. –/65 1T. 45.8 t.

170 301	**FT**	P	*TP*	XW	50301	79301
170 302	**FT**	P	*TP*	XW	50302	79302
170 303	**FT**	P	*TP*	XW	50303	79303
170 304	**FT**	P	*TP*	XW	50304	79304
170 305	**FT**	P	*TP*	XW	50305	79305
170 306	**FT**	P	*TP*	XW	50306	79306
170 307	**FT**	P	*TP*	XW	50307	79307
170 308	**FT**	P	*TP*	XW	50308	79308
170 309	**FT**	P	*TP*	XW	50399	79399

Class 170/3. Units built for Hull Trains, now in use with ScotRail. Chapman seating. DMSL–MSLRB–DMSL.

DMSL(A). Bombardier Derby 2004. –/55 1TD 2W. 46.5 t.
MSLRB. Bombardier Derby 2004. –/57 1T. Buffet and guard's office 44.7 t.
DMSL(B). Bombardier Derby 2004. –/67 1T. 47.0 t.

170 393	**SR**	P	*SR*	HA	50393	56393	79393
170 394	**SR**	P	*SR*	HA	50394	56394	79394
170 395	**SR**	P	*SR*	HA	50395	56395	79395
170 396	**SR**	P	*SR*	HA	50396	56396	79396

Class 170/3. CrossCountry units. Lazareni seating. DMSL–MS–DMCL.

DMSL. Bombardier Derby 2002. –/59 1TD 2W. 45.4 t.
MS. Bombardier Derby 2002. –/80. 43.0 t.
DMCL. Bombardier Derby 2002. 9/52 1T. 45.8 t.

| 170 397 | **XC** | P | *XC* | TS | 50397 | 56397 | 79397 |
| 170 398 | **XC** | P | *XC* | TS | 50398 | 56398 | 79398 |

Class 170/4. ScotRail "express" units. Chapman seating. DMCL–MS–DMCL.

DMCL(A). Adtranz Derby 1999–2001. 9/43 1TD 2W. 45.2 t.
MS. Adtranz Derby 1999–2001. –/76. 42.5 t.
DMCL(B). Adtranz Derby 1999–2001. 9/49 1T. 45.2 t.

170 401	**FS**	P	*SR*	HA	50401	56401	79401
170 402	**SR**	P	*SR*	HA	50402	56402	79402
170 403	**FS**	P	*SR*	HA	50403	56403	79403
170 404	**FS**	P	*SR*	HA	50404	56404	79404
170 405	**FS**	P	*SR*	HA	50405	56405	79405
170 406	**FS**	P	*SR*	HA	50406	56406	79406
170 407	**FS**	P	*SR*	HA	50407	56407	79407
170 408	**FS**	P	*SR*	HA	50408	56408	79408
170 409	**FS**	P	*SR*	HA	50409	56409	79409
170 410	**FS**	P	*SR*	HA	50410	56410	79410
170 411	**FS**	P	*SR*	HA	50411	56411	79411
170 412	**SR**	P	*SR*	HA	50412	56412	79412
170 413	**FS**	P	*SR*	HA	50413	56413	79413
170 414	**FS**	P	*SR*	HA	50414	56414	79414
170 415	**SR**	P	*SR*	HA	50415	56415	79415
170 416	**FS**	E	*SR*	HA	50416	56416	79416
170 417	**FS**	E	*SR*	HA	50417	56417	79417
170 418	**SR**	E	*SR*	HA	50418	56418	79418

170 419	**FS**	E	*SR*	HA	50419	56419	79419
170 420	**FS**	E	*SR*	HA	50420	56420	79420
170 421	**FS**	E	*SR*	HA	50421	56421	79421
170 422	**FS**	E	*SR*	HA	50422	56422	79422
170 423	**FS**	E	*SR*	HA	50423	56423	79423
170 424	**FS**	E	*SR*	HA	50424	56424	79424

Names (carried on end cars):

170 401	Sir Moir Lockhead OBE
170 405	Riverside Museum
170 407	UNIVERSITY OF ABERDEEN

Class 170/4. ScotRail "express" units. Chapman seating. DMCL–MS–DMCL.

DMCL. Bombardier Derby 2003–05. 9/43 1TD 2W. 46.8 t.
MS. Bombardier Derby 2003–05. –/76. 43.7 t.
DMCL. Bombardier Derby 2003–05. 9/49 1T. 46.5 t.

170 425	**SR**	P	*SR*	HA	50425	56425	79425
170 426	**SR**	P	*SR*	HA	50426	56426	79426
170 427	**SR**	P	*SR*	HA	50427	56427	79427
170 428	**SR**	P	*SR*	HA	50428	56428	79428
170 429	**SR**	P	*SR*	HA	50429	56429	79429
170 430	**SR**	P	*SR*	HA	50430	56430	79430
170 431	**SR**	P	*SR*	HA	50431	56431	79431
170 432	**SR**	P	*SR*	HA	50432	56432	79432
170 433	**SR**	P	*SR*	HA	50433	56433	79433
170 434	**SR**	P	*SR*	HA	50434	56434	79434

Class 170/4. ScotRail units. Originally built as Standard Class only units.
170 450–457 have been retro-fitted with First Class. Chapman seating.
DMSL–MS–DMSL or † DMCL–MS–DMCL.

DMSL. Bombardier Derby 2004–05. –/55 1TD 2W († 9/47 1TD 2W). 46.3 t.
MS. Bombardier Derby 2004–05. –/76. 43.4 t.
DMSL. Bombardier Derby 2004–05. –/67 1T († 9/49 1T 1W). 46.4 t.

170 450	†	**SR**	P	*SR*	HA	50450	56450	79450
170 451	†	**SR**	P	*SR*	HA	50451	56451	79451
170 452	†	**SR**	P	*SR*	HA	50452	56452	79452
170 453	†	**SR**	P	*SR*	HA	50453	56453	79453
170 454	†	**SR**	P	*SR*	HA	50454	56454	79454
170 455	†	**SR**	P	*SR*	HA	50455	56455	79455
170 456	†	**SR**	P	*SR*	HA	50456	56456	79456
170 457	†	**SR**	P	*SR*	HA	50457	56457	79457
170 458		**SR**	P	*SR*	HA	50458	56458	79458
170 459		**SR**	P	*SR*	HA	50459	56459	79459
170 460		**SR**	P	*SR*	HA	50460	56460	79460
170 461		**SR**	P	*SR*	HA	50461	56461	79461

Class 170/4. ScotRail units. Standard Class only units. Chapman seating.
DMSL–MS–DMSL.

50470–471. DMSL(A). Adtranz Derby 2001. –/55 1TD 2W. 45.1 t.
50472–478. DMSL(A). Bombardier Derby 2004–05. –/57 1TD 2W. 46.3 t.

56470–471. MS. Adtranz Derby 2001. –/76. 42.4 t.
56472–478. MS. Bombardier Derby 2004–05. –/76. 43.4 t.
79470–471. DMSL(B). Adtranz Derby 2001. –/67 1T. 45.1 t.
79472–478. DMSL(B). Bombardier Derby 2004–05. –/67 1T. 46.4 t.

170 470	**SR**	P	*SR*	HA	50470	56470	79470
170 471	**SR**	P	*SR*	HA	50471	56471	79471
170 472	**SR**	P	*SR*	HA	50472	56472	79472
170 473	**SR**	P	*SR*	HA	50473	56473	79473
170 474	**SR**	P	*SR*	HA	50474	56474	79474
170 475	**SR**	P	*SR*	HA	50475	56475	79475
170 476	**SR**	P	*SR*	HA	50476	56476	79476
170 477	**SR**	P	*SR*	HA	50477	56477	79477
170 478	**SR**	P	*SR*	HA	50478	56478	79478

Class 170/5. London Midland and CrossCountry 2-car units. Lazareni seating. DMSL–DMSL or * DMSL–DMCL (CrossCountry).

DMSL(A). Adtranz Derby 1999–2000. –/55 1TD 2W (* –/59 1TD 2W). 45.8 t.
DMSL(B). Adtranz Derby 1999–2000. –/67 1T (* DMCL 9/52 1T). 45.9 t.

170 501		**LM**	P	*LM*	TS	50501	79501
170 502		**LM**	P	*LM*	TS	50502	79502
170 503		**LM**	P	*LM*	TS	50503	79503
170 504		**LM**	P	*LM*	TS	50504	79504
170 505		**LM**	P	*LM*	TS	50505	79505
170 506		**LM**	P	*LM*	TS	50506	79506
170 507		**LM**	P	*LM*	TS	50507	79507
170 508		**LM**	P	*LM*	TS	50508	79508
170 509		**LM**	P	*LM*	TS	50509	79509
170 510		**LM**	P	*LM*	TS	50510	79510
170 511		**LM**	P	*LM*	TS	50511	79511
170 512		**LM**	P	*LM*	TS	50512	79512
170 513		**LM**	P	*LM*	TS	50513	79513
170 514		**LM**	P	*LM*	TS	50514	79514
170 515		**LM**	P	*LM*	TS	50515	79515
170 516		**LM**	P	*LM*	TS	50516	79516
170 517		**LM**	P	*LM*	TS	50517	79517
170 518	*	**XC**	P	*XC*	TS	50518	79518
170 519	*	**XC**	P	*XC*	TS	50519	79519
170 520	*	**XC**	P	*XC*	TS	50520	79520
170 521	*	**XC**	P	*XC*	TS	50521	79521
170 522	*	**XC**	P	*XC*	TS	50522	79522
170 523	*	**XC**	P	*XC*	TS	50523	79523

Class 170/6. London Midland and CrossCountry 3-car units. Lazareni seating. DMSL–MS–DMSL or * DMSL–MS–DMCL (CrossCountry).

DMSL(A). Adtranz Derby 2000. –/55 1TD 2W (* –/59 1TD 2W). 45.8 t.
MS. Adtranz Derby 2000. –/74 (* –/80). 42.4 t.
DMSL(B). Adtranz Derby 2000. –/67 1T (* DMCL 9/52 1T). 45.9 t.

170 630	**LM**	P	*LM*	TS	50630	56630	79630
170 631	**LM**	P	*LM*	TS	50631	56631	79631

170 632	**LM**	P	*LM*	TS	50632	56632	79632
170 633	**LM**	P	*LM*	TS	50633	56633	79633
170 634	**LM**	P	*LM*	TS	50634	56634	79634
170 635	**LM**	P	*LM*	TS	50635	56635	79635
170 636	*	**XC**	P	*XC*	TS	50636	56636	79636
170 637	*	**XC**	P	*XC*	TS	50637	56637	79637
170 638	*	**XC**	P	*XC*	TS	50638	56638	79638
170 639	*	**XC**	P	*XC*	TS	50639	56639	79639

CLASS 171   TURBOSTAR   BOMBARDIER

DMCL–DMSL or DMCL–MS–MS–DMCL. Southern units. Air conditioned. Chapman seating.

Construction: Welded aluminium bodies with bolt-on steel ends.
Engines: One MTU 6R183TD13H of 315 kW (422 hp) at 1900 rpm.
Transmission: Hydraulic. Voith T211rzze to ZF final drive.
Bogies: One Adtranz P3–23 and one BREL T3–23 per car.
Couplers: Dellner 12 at outer ends, bar within unit (Class 171/8s).
Dimensions: 23.62/23.61 x 2.69 m.
Gangways: Within unit only.   **Wheel Arrangement:** 2-B (+ B-2 + B-2) + B-2.
Doors: Twin-leaf swing plug.   **Maximum Speed:** 100 mph.
Seating Layout: 1: 2+1 facing/unidirectional. 2: 2+2 facing/unidirectional.
Multiple Working: Within class and with EMU Classes 375 and 377 in an emergency.

Class 171/7. 2-car units. DMCL–DMSL.

171 721–726 were built as Class 170s (170 721–726), but renumbered as Class 171s on fitting with Dellner couplers.

171 730 was formerly South West Trains unit 170 392, before transferring to Southern in 2007.

50721–726. DMCL. Bombardier Derby 2003. 9/43 1TD 2W. 47.6 t.
50727–729. DMCL. Bombardier Derby 2005. 9/43 1TD 2W. 46.3 t.
50392. DMCL. Bombardier Derby 2003. 9/43 1TD 2W. 46.6 t.
79721–726. DMSL. Bombardier Derby 2003. –/64 1T. 47.8 t.
79727–729. DMSL. Bombardier Derby 2005. –/64 1T. 46.2 t.
79392. DMSL. Bombardier Derby 2003. –/64 1T. 46.5 t.

171 721	**SN**	P	*SN*	SU	50721	79721
171 722	**SN**	P	*SN*	SU	50722	79722
171 723	**SN**	P	*SN*	SU	50723	79723
171 724	**SN**	P	*SN*	SU	50724	79724
171 725	**SN**	P	*SN*	SU	50725	79725
171 726	**SN**	P	*SN*	SU	50726	79726
171 727	**SN**	P	*SN*	SU	50727	79727
171 728	**SN**	P	*SN*	SU	50728	79728
171 729	**SN**	P	*SN*	SU	50729	79729
171 730	**SN**	P	*SN*	SU	50392	79392

Class 171/8. 4-car units. DMCL(A)–MS–MS–DMCL(B).

DMCL(A). Bombardier Derby 2004. 9/43 1TD 2W. 46.5 t.
MS. Bombardier Derby 2004. –/74. 43.7 t.
DMCL(B). Bombardier Derby 2004. 9/50 1T. 46.5 t.

171 801	**SN**	P	*SN*	SU	50801	54801	56801	79801
171 802	**SN**	P	*SN*	SU	50802	54802	56802	79802
171 803	**SN**	P	*SN*	SU	50803	54803	56803	79803
171 804	**SN**	P	*SN*	SU	50804	54804	56804	79804
171 805	**SN**	P	*SN*	SU	50805	54805	56805	79805
171 806	**SN**	P	*SN*	SU	50806	54806	56806	79806

CLASS 172   TURBOSTAR   BOMBARDIER

New generation London Overground, Chiltern Railways and London Midland Turbostars. Air conditioned.

Construction: Welded aluminium bodies with bolt-on steel ends.
Engines: One MTU 6H1800R83 of 360 kW (483 hp) at 1800 rpm.
Transmission: Mechanical. Supplied by ZF, Germany.
Bogies: B5006 type "lightweight" bogies.
Couplers: BSI at outer ends, bar within unit.
Dimensions: 23.62/23.0 x 2.69 m.
Gangways: London Overground & Chiltern units: Within unit only. London Midland units: Throughout.
Wheel Arrangement: 2-B (+ B-2) + B-2.
Doors: Twin-leaf sliding plug.
Maximum Speed: 100 mph (London Overground units 75 mph).
Seating Layout: 2+2 facing/unidirectional.
Multiple Working: Within class and with Classes 150, 153, 155, 156, 158, 159, 165, 166 and 170.

Class 172/0. London Overground units. Used on the Gospel Oak–Barking line. DMS–DMS.

59311–318. DMS(W). Bombardier Derby 2009–10. –/60 2W. 41.6 t.
59411–418. DMS. Bombardier Derby 2009–10. –/64. 41.5 t.

172 001	**LO**	A	*LO*	WN	59311	59411
172 002	**LO**	A	*LO*	WN	59312	59412
172 003	**LO**	A	*LO*	WN	59313	59413
172 004	**LO**	A	*LO*	WN	59314	59414
172 005	**LO**	A	*LO*	WN	59315	59415
172 006	**LO**	A	*LO*	WN	59316	59416
172 007	**LO**	A	*LO*	WN	59317	59417
172 008	**LO**	A	*LO*	WN	59318	59418

Class 172/1. Chiltern Railways units. DMSL–DMS.

59111–114. DMSL. Bombardier Derby 2009–10. –/60(5) 1TD 2W. 42.4 t.
59211–214. DMS. Bombardier Derby 2009–10. –/80. 41.8 t.

172 101	**CR**	A	*CR*	AL	59111	59211
172 102	**CR**	A	*CR*	AL	59112	59212

| 172 103 | **CR** | A | *CR* | AL | 59113 | 59213 |
| 172 104 | **CR** | A | *CR* | AL | 59114 | 59214 |

Class 172/2. London Midland 2-car units. DMSL–DMS. Used on local services via Birmingham Snow Hill.

50211–222. DMSL. Bombardier Derby 2010–11. –/52(11) 1TD 2W. 42.5 t.
79211–222. DMS. Bombardier Derby 2010–11. –/68(8). 41.9 t.

172 211	**LM**	P	*LM*	TS	50211	79211
172 212	**LM**	P	*LM*	TS	50212	79212
172 213	**LM**	P	*LM*	TS	50213	79213
172 214	**LM**	P	*LM*	TS	50214	79214
172 215	**LM**	P	*LM*	TS	50215	79215
172 216	**LM**	P	*LM*	TS	50216	79216
172 217	**LM**	P	*LM*	TS	50217	79217
172 218	**LM**	P	*LM*	TS	50218	79218
172 219	**LM**	P	*LM*	TS	50219	79219
172 220	**LM**	P	*LM*	TS	50220	79220
172 221	**LM**	P	*LM*	TS	50221	79221
172 222	**LM**	P	*LM*	TS	50222	79222

Class 172/3. London Midland 3-car units. DMSL–MS–DMS. Used on local services via Birmingham Snow Hill.

50331–345. DMSL. Bombardier Derby 2010–11. –/52(11) 1TD 2W. 42.5 t.
56331–345. MS. Bombardier Derby 2010–11. –/72(8). 38.8 t.
79331–345. DMS. Bombardier Derby 2010–11. –/68(8). 41.9 t.

172 331	**LM**	P	*LM*	TS	50331	56331	79331
172 332	**LM**	P	*LM*	TS	50332	56332	79332
172 333	**LM**	P	*LM*	TS	50333	56333	79333
172 334	**LM**	P	*LM*	TS	50334	56334	79334
172 335	**LM**	P	*LM*	TS	50335	56335	79335
172 336	**LM**	P	*LM*	TS	50336	56336	79336
172 337	**LM**	P	*LM*	TS	50337	56337	79337
172 338	**LM**	P	*LM*	TS	50338	56338	79338
172 339	**LM**	P	*LM*	TS	50339	56339	79339
172 340	**LM**	P	*LM*	TS	50340	56340	79340
172 341	**LM**	P	*LM*	TS	50341	56341	79341
172 342	**LM**	P	*LM*	TS	50342	56342	79342
172 343	**LM**	P	*LM*	TS	50343	56343	79343
172 344	**LM**	P	*LM*	TS	50344	56344	79344
172 345	**LM**	P	*LM*	TS	50345	56345	79345

CLASS 175   CORADIA 1000   ALSTOM

Air conditioned.

Construction: Steel.
Engines: One Cummins N14 of 335 kW (450 hp).
Transmission: Hydraulic. Voith T211rzze to ZF Voith final drive.
Bogies: ACR (Alstom FBO) – LTB-MBS1, TB-MB1, MBS1-LTB.
Couplers: Scharfenberg outer ends and bar within unit (Class 175/1).

Dimensions: 23.7 x 2.73 m.
Gangways: Within unit only.   **Wheel Arrangement:** 2-B (+ B-2) + B-2.
Doors: Single-leaf swing plug.   **Maximum Speed:** 100 mph.
Seating Layout: 2+2 facing/unidirectional.
Multiple Working: Within class and with Class 180.

Class 175/0. DMSL–DMSL. 2-car units.

DMSL(A). Alstom Birmingham 1999–2000. –/54 1TD 2W. 48.8 t.
DMSL(B). Alstom Birmingham 1999–2000. –/64 1T. 50.7 t.

175 001	**AV**	A	*AW*	CH	50701	79701
175 002	**AV**	A	*AW*	CH	50702	79702
175 003	**AV**	A	*AW*	CH	50703	79703
175 004	**AV**	A	*AW*	CH	50704	79704
175 005	**AV**	A	*AW*	CH	50705	79705
175 006	**AV**	A	*AW*	CH	50706	79706
175 007	**AV**	A	*AW*	CH	50707	79707
175 008	**AV**	A	*AW*	CH	50708	79708
175 009	**AV**	A	*AW*	CH	50709	79709
175 010	**AV**	A	*AW*	CH	50710	79710
175 011	**AV**	A	*AW*	CH	50711	79711

Class 175/1. DMSL–MSL–DMSL. 3-car units.

DMSL(A). Alstom Birmingham 1999–2001. –/54 1TD 2W. 50.7 t.
MSL. Alstom Birmingham 1999–2001. –/68 1T. 47.5 t.
DMSL(B). Alstom Birmingham 1999–2001. –/64 1T. 49.5 t.

175 101	**AV**	A	*AW*	CH	50751	56751	79751
175 102	**AV**	A	*AW*	CH	50752	56752	79752
175 103	**AV**	A	*AW*	CH	50753	56753	79753
175 104	**AV**	A	*AW*	CH	50754	56754	79754
175 105	**AV**	A	*AW*	CH	50755	56755	79755
175 106	**AV**	A	*AW*	CH	50756	56756	79756
175 107	**AV**	A	*AW*	CH	50757	56757	79757
175 108	**AV**	A	*AW*	CH	50758	56758	79758
175 109	**AV**	A	*AW*	CH	50759	56759	79759
175 110	**AV**	A	*AW*	CH	50760	56760	79760
175 111	**AV**	A	*AW*	CH	50761	56761	79761
175 112	**AV**	A	*AW*	CH	50762	56762	79762
175 113	**AV**	A	*AW*	CH	50763	56763	79763
175 114	**AV**	A	*AW*	CH	50764	56764	79764
175 115	**AV**	A	*AW*	CH	50765	56765	79765
175 116	**AV**	A	*AW*	CH	50766	56766	79766

CLASS 180   ADELANTE   ALSTOM

Air conditioned.

Construction: Steel.
Engines: One Cummins QSK19 of 560 kW (750 hp) at 2100 rpm.
Transmission: Hydraulic. Voith T312br to Voith final drive.
Bogies: ACR (Alstom FBO): LTB1-MBS2, TB1-MB2, TB1-MB2, TB2-MB2, MBS2-LTB1.

Couplers: Scharfenberg outer ends, bar within unit.
Dimensions: 23.71/23.03 x 2.73 m.
Gangways: Within unit only.
Wheel Arrangement: 2-B + B-2 + B-2 + B-2 + B-2.
Doors: Single-leaf swing plug.   **Maximum Speed:** 125 mph.
Seating Layout: 1: 2+1 facing/unidirectional, 2: 2+2 facing/unidirectional.
Multiple Working: Within class and with Class 175.

DMSL(A). Alstom Birmingham 2000–01. –/46 2W 1TD. 51.7 t.
MFL. Alstom Birmingham 2000–01. 42/– 1T 1W + catering point. 49.6 t.
MSL. Alstom Birmingham 2000–01. –/68 1T. 49.5 t.
MSLRB. Alstom Birmingham 2000–01. –/56 1T. 50.3 t.
DMSL(B). Alstom Birmingham 2000–01. –/56 1T. 51.4 t.

180 101	**GC**	A	*GC*	HT	50901	54901	55901	56901	59901
180 102	**FD**	A	*GW*	OO	50902	54902	55902	56902	59902
180 103	**FD**	A	*GW*	OO	50903	54903	55903	56903	59903
180 104	**FD**	A	*GW*	OO	50904	54904	55904	56904	59904
180 105	**GC**	A	*GC*	HT	50905	54905	55905	56905	59905
180 106	**FD**	A	*GW*	OO	50906	54906	55906	56906	59906
180 107	**GC**	A	*GC*	HT	50907	54907	55907	56907	59907
180 108	**FD**	A	*GW*	OO	50908	54908	55908	56908	59908
180 109	**FD**	A	*HT*	OO	50909	54909	55909	56909	59909
180 110	**FD**	A	*HT*	OO	50910	54910	55910	56910	59910
180 111	**FD**	A	*HT*	OO	50911	54911	55911	56911	59911
180 112	**GC**	A	*GC*	HT	50912	54912	55912	56912	59912
180 113	**FD**	A	*HT*	OO	50913	54913	55913	56913	59913
180 114	**GC**	A	*GC*	HT	50914	54914	55914	56914	59914

Names (carried on DMSL(A):

180 105	THE YORKSHIRE ARTIST ASHLEY JACKSON
180 107	HART OF THE NORTH
180 112	JAMES HERRIOT

CLASS 185   DESIRO UK   SIEMENS

Air conditioned. Grammer seating.

Construction: Aluminium.
Engines: One Cummins QSK19 of 560 kW (750 hp) at 2100 rpm.
Transmission: Voith.
Bogies: Siemens.
Couplers: Dellner 12.   **Dimensions:** 23.76/23.75 x 2.66 m.
Gangways: Within unit only.   **Wheel Arrangement:** 2-B + 2-B + B-2.
Doors: Double-leaf sliding plug.   **Maximum Speed:** 100 mph.
Seating Layout: 1: 2+1 facing/unidirectional, 2: 2+2 facing/unidirectional.
Multiple Working: Within class only.

DMCL. Siemens Krefeld 2005–06. 15/18(8) 2W 1TD + catering point. 55.4 t.
MSL. Siemens Krefeld 2005–06. –/72 1T. 52.7 t.
DMS. Siemens Krefeld 2005–06. –/64(4). 54.9 t.

185 101	**FT**	E	*TP*	AK	51101	53101	54101
185 102	**FT**	E	*TP*	AK	51102	53102	54102
185 103	**FT**	E	*TP*	AK	51103	53103	54103
185 104	**FT**	E	*TP*	AK	51104	53104	54104
185 105	**FT**	E	*TP*	AK	51105	53105	54105
185 106	**FT**	E	*TP*	AK	51106	53106	54106
185 107	**FT**	E	*TP*	AK	51107	53107	54107
185 108	**FT**	E	*TP*	AK	51108	53108	54108
185 109	**FT**	E	*TP*	AK	51109	53109	54109
185 110	**FT**	E	*TP*	AK	51110	53110	54110
185 111	**FT**	E	*TP*	AK	51111	53111	54111
185 112	**FT**	E	*TP*	AK	51112	53112	54112
185 113	**FT**	E	*TP*	AK	51113	53113	54113
185 114	**FT**	E	*TP*	AK	51114	53114	54114
185 115	**FT**	E	*TP*	AK	51115	53115	54115
185 116	**FT**	E	*TP*	AK	51116	53116	54116
185 117	**FT**	E	*TP*	AK	51117	53117	54117
185 118	**FT**	E	*TP*	AK	51118	53118	54118
185 119	**FT**	E	*TP*	AK	51119	53119	54119
185 120	**FT**	E	*TP*	AK	51120	53120	54120
185 121	**FT**	E	*TP*	AK	51121	53121	54121
185 122	**FT**	E	*TP*	AK	51122	53122	54122
185 123	**FT**	E	*TP*	AK	51123	53123	54123
185 124	**FT**	E	*TP*	AK	51124	53124	54124
185 125	**FT**	E	*TP*	AK	51125	53125	54125
185 126	**FT**	E	*TP*	AK	51126	53126	54126
185 127	**FT**	E	*TP*	AK	51127	53127	54127
185 128	**FT**	E	*TP*	AK	51128	53128	54128
185 129	**FT**	E	*TP*	AK	51129	53129	54129
185 130	**FT**	E	*TP*	AK	51130	53130	54130
185 131	**FT**	E	*TP*	AK	51131	53131	54131
185 132	**FT**	E	*TP*	AK	51132	53132	54132
185 133	**FT**	E	*TP*	AK	51133	53133	54133
185 134	**FT**	E	*TP*	AK	51134	53134	54134
185 135	**FT**	E	*TP*	AK	51135	53135	54135
185 136	**FT**	E	*TP*	AK	51136	53136	54136
185 137	**FT**	E	*TP*	AK	51137	53137	54137
185 138	**FT**	E	*TP*	AK	51138	53138	54138
185 139	**FT**	E	*TP*	AK	51139	53139	54139
185 140	**FT**	E	*TP*	AK	51140	53140	54140
185 141	**FT**	E	*TP*	AK	51141	53141	54141
185 142	**FT**	E	*TP*	AK	51142	53142	54142
185 143	**FT**	E	*TP*	AK	51143	53143	54143
185 144	**FT**	E	*TP*	AK	51144	53144	54144
185 145	**FT**	E	*TP*	AK	51145	53145	54145
185 146	**FT**	E	*TP*	AK	51146	53146	54146
185 147	**FT**	E	*TP*	AK	51147	53147	54147
185 148	**FT**	E	*TP*	AK	51148	53148	54148
185 149	**FT**	E	*TP*	AK	51149	53149	54149
185 150	**FT**	E	*TP*	AK	51150	53150	54150
185 151	**FT**	E	*TP*	AK	51151	53151	54151

2. DIESEL ELECTRIC UNITS

CLASS 201/202   PRESERVED "HASTINGS" UNIT   BR

DMBS–TSL–TSL–TSRB–TSL–DMBS.

Preserved unit made up from two Class 201 short-frame cars and three Class 202 long-frame cars. The "Hastings" units were made with narrow body-profiles for use on the section between Tonbridge and Battle which had tunnels of restricted loading gauge. These tunnels were converted to single track operation in the 1980s thus allowing standard loading gauge stock to be used. The set also contains a Class 411 EMU trailer (not Hastings line gauge) and a Class 422 EMU buffet car.

Construction: Steel.
Engine: One English Electric 4SRKT Mk. 2 of 450 kW (600 hp) at 850 rpm.
Main Generator: English Electric EE824.
Traction Motors: Two English Electric EE507 mounted on the inner bogie.
Bogies: SR Mk 4. (Former EMU TSL vehicles have Commonwealth bogies).
Couplers: Drophead buckeye.
Dimensions: 18.40 x 2.50 m (60000), 20.35 x 2.50 m (60116/118/529), 18.36 x 2.50 m (60501), 20.35 x 2.82 m (69337), 20.30 x 2.82 m (70262).
Gangways: Within unit only.   **Doors:** Manually operated slam.
Brakes: Electro-pneumatic and automatic air.
Maximum Speed: 75 mph.   **Seating Layout:** 2+2 facing.
Multiple Working: Other ex-BR Southern Region DEMU vehicles.

60000. DMBS. Lot No. 30329 Eastleigh 1957. –/22. 55.0 t.
60116. DMBS. Lot No. 30395 Eastleigh 1957. –/31. 56.0 t.
60118. DMBS. Lot No. 30395 Eastleigh 1957. –/30. 56.0 t.
60501. TSL. Lot No. 30331 Eastleigh 1957. –/52 2T. 29.5 t.
60529. TSL. Lot No. 30397 Eastleigh 1957. –/60 2T. 30.5 t.
69337. TSRB (ex-Class 422 EMU). Lot No. 30805 York 1970. –/40. 35.0 t.
70262. TSL (ex-Class 411/5 EMU). Lot No. 30455 Eastleigh 1958. –/64 2T. 31.5 t.

201 001	G	HD	*HD*	SE	60116	60529	70262	69337	60501	60118
Spare	G	HD	*HD*	SE	60000					

Names:

60000	Hastings
60116	Mountfield
60118	Tunbridge Wells

CLASS 220   VOYAGER   BOMBARDIER

DMS–MS–MS–DMF. All engines have been derated from 750 hp to 700 hp.

Construction: Steel.
Engine: Cummins QSK19 of 520 kW (700 hp) at 1800 rpm.
Transmission: Two Alstom Onix 800 three-phase traction motors of 275 kW.
Braking: Rheostatic and electro-pneumatic.

Bogies: Bombardier B5005.
Couplers: Dellner 12 at outer ends, bar within unit.
Dimensions: 23.85/23.00 (602xx) x 2.73 m.
Gangways: Within unit only.
Wheel Arrangement: 1A-A1 + 1A-A1 + 1A-A1 + 1A-A1.
Doors: Single-leaf swing plug.
Maximum Speed: 125 m.p.h.
Seating Layout: 1: 2+1 facing/unidirectional, 2: 2+2 mainly unidirectional.
Multiple Working: Within class and with Classes 221 and 222 (in an emergency). Also can be controlled from Class 57/3 locomotives.

DMS. Bombardier Bruges/Wakefield 2000–01. –/42 1TD 1W. 51.1 t.
MS(A). Bombardier Bruges/Wakefield 2000–01. –/66. 45.9 t.
MS(B). Bombardier Bruges/Wakefield 2000–01. –/66 1TD. 46.7 t.
DMF. Bombardier Bruges/Wakefield 2000–01. 26/– 1TD 1W. 50.9 t.

220 001	**XC**	VL	*XC*	CZ	60301	60701	60201	60401
220 002	**XC**	VL	*XC*	CZ	60302	60702	60202	60402
220 003	**XC**	VL	*XC*	CZ	60303	60703	60203	60403
220 004	**XC**	VL	*XC*	CZ	60304	60704	60204	60404
220 005	**XC**	VL	*XC*	CZ	60305	60705	60205	60405
220 006	**XC**	VL	*XC*	CZ	60306	60706	60206	60406
220 007	**XC**	VL	*XC*	CZ	60307	60707	60207	60407
220 008	**XC**	VL	*XC*	CZ	60308	60708	60208	60408
220 009	**XC**	VL	*XC*	CZ	60309	60709	60209	60409
220 010	**XC**	VL	*XC*	CZ	60310	60710	60210	60410
220 011	**XC**	VL	*XC*	CZ	60311	60711	60211	60411
220 012	**XC**	VL	*XC*	CZ	60312	60712	60212	60412
220 013	**XC**	VL	*XC*	CZ	60313	60713	60213	60413
220 014	**XC**	VL	*XC*	CZ	60314	60714	60214	60414
220 015	**XC**	VL	*XC*	CZ	60315	60715	60215	60415
220 016	**XC**	VL	*XC*	CZ	60316	60716	60216	60416
220 017	**XC**	VL	*XC*	CZ	60317	60717	60217	60417
220 018	**XC**	VL	*XC*	CZ	60318	60718	60218	60418
220 019	**XC**	VL	*XC*	CZ	60319	60719	60219	60419
220 020	**XC**	VL	*XC*	CZ	60320	60720	60220	60420
220 021	**XC**	VL	*XC*	CZ	60321	60721	60221	60421
220 022	**XC**	VL	*XC*	CZ	60322	60722	60222	60422
220 023	**XC**	VL	*XC*	CZ	60323	60723	60223	60423
220 024	**XC**	VL	*XC*	CZ	60324	60724	60224	60424
220 025	**XC**	VL	*XC*	CZ	60325	60725	60225	60425
220 026	**XC**	VL	*XC*	CZ	60326	60726	60226	60426
220 027	**XC**	VL	*XC*	CZ	60327	60727	60227	60427
220 028	**XC**	VL	*XC*	CZ	60328	60728	60228	60428
220 029	**XC**	VL	*XC*	CZ	60329	60729	60229	60429
220 030	**XC**	VL	*XC*	CZ	60330	60730	60230	60430
220 031	**XC**	VL	*XC*	CZ	60331	60731	60231	60431
220 032	**XC**	VL	*XC*	CZ	60332	60732	60232	60432
220 033	**XC**	VL	*XC*	CZ	60333	60733	60233	60433
220 034	**XC**	VL	*XC*	CZ	60334	60734	60234	60434

CLASS 221   SUPER VOYAGER   BOMBARDIER

* DMS–MS–MS–MSRMB–DMF (Virgin Trains units) or DMS–MS–MS–MS–DMF (CrossCountry units). Built as tilting units but tilt now isolated on CrossCountry sets. All engines have been derated from 750 hp to 700 hp.

Construction: Steel.
Engine: Cummins QSK19 of 520 kW (700 hp) at 1800 rpm.
Transmission: Two Alstom Onix 800 three-phase traction motors of 275 kW.
Braking: Rheostatic and electro-pneumatic.
Bogies: Bombardier HVP.
Couplers: Dellner 12 at outer ends, bar within unit.
Dimensions: 23.67 x 2.73 m.
Gangways: Within unit only.
Wheel Arrangement: 1A-A1 + 1A-A1 + 1A-A1 (+ 1A-A1) + 1A-A1.
Doors: Single-leaf swing plug.
Maximum Speed: 125 mph.
Seating Layout: 1: 2+1 facing/unidirectional, 2: 2+2 mainly unidirectional.
Multiple Working: Within class and with Classes 220 and 222 (in an emergency). Also can be controlled from Class 57/3 locomotives.

* Virgin Trains units. MSRMB moved adjacent to the DMF. The seating in this vehicle (2+2 facing) can be used by First or Standard Class passengers depending on demand.

Advertising livery: 221 115 Dark grey Bombardier branding on end vehicles.

DMS. Bombardier Bruges/Wakefield 2001–02. –/42 1TD 1W. 58.5 t (* 58.9 t.)
60751–794 MS (* MSRMB). Bombardier Bruges/Wakefield 2001–02. –/66 (* –/52). 54.1 t (* 55.9 t.)
60951–994. MS. Bombardier Bruges/Wakefield 2001–02. –/66 1TD (* –/68 1TD). 54.8 t (* 54.3 t.)
60851–890. MS. Bombardier Bruges/Wakefield 2001–02. –/62 1TD (* –/68 1TD). 54.4 t (* 55.0 t.)
DMF. Bombardier Bruges/Wakefield 2001–02. 26/– 1TD 1W. 58.9 t (* 59.1 t.)

221 101	*	**VT**	VL	*VW*	CZ	60351	60951	60851	60751	60451
221 102	*	**VT**	VL	*VW*	CZ	60352	60952	60852	60752	60452
221 103	*	**VT**	VL	*VW*	CZ	60353	60953	60853	60753	60453
221 104	*	**VT**	VL	*VW*	CZ	60354	60954	60854	60754	60454
221 105	*	**VT**	VL	*VW*	CZ	60355	60955	60855	60755	60455
221 106	*	**VT**	VL	*VW*	CZ	60356	60956	60856	60756	60456
221 107	*	**VT**	VL	*VW*	CZ	60357	60957	60857	60757	60457
221 108	*	**VT**	VL	*VW*	CZ	60358	60958	60858	60758	60458
221 109	*	**VT**	VL	*VW*	CZ	60359	60959	60859	60759	60459
221 110	*	**VT**	VL	*VW*	CZ	60360	60960	60860	60760	60460
221 111	*	**VT**	VL	*VW*	CZ	60361	60961	60861	60761	60461
221 112	*	**VT**	VL	*VW*	CZ	60362	60962	60862	60762	60462
221 113	*	**VT**	VL	*VW*	CZ	60363	60963	60863	60763	60463
221 114	*	**VT**	VL	*VW*	CZ	60364	60764	60964	60864	60464
221 115	*	**AL**	VL	*VW*	CZ	60365	60765	60965	60865	60465
221 116	*	**VT**	VL	*VW*	CZ	60366	60766	60966	60866	60466
221 117	*	**VT**	VL	*VW*	CZ	60367	60767	60967	60867	60467

221 118	*	**VT**	VL	*VW*	CZ	60368	60768	60968	60868	60468
221 119		**XC**	VL	*XC*	CZ	60369	60769	60969	60869	60469
221 120		**XC**	VL	*XC*	CZ	60370	60770	60970	60870	60470
221 121		**XC**	VL	*XC*	CZ	60371	60771	60971	60871	60471
221 122		**XC**	VL	*XC*	CZ	60372	60772	60972	60872	60472
221 123		**XC**	VL	*XC*	CZ	60373	60773	60973	60873	60473
221 124		**XC**	VL	*XC*	CZ	60374	60774	60974	60874	60474
221 125		**XC**	VL	*XC*	CZ	60375	60775	60975	60875	60475
221 126		**XC**	VL	*XC*	CZ	60376	60776	60976	60876	60476
221 127		**XC**	VL	*XC*	CZ	60377	60777	60977	60877	60477
221 128		**XC**	VL	*XC*	CZ	60378	60778	60978	60878	60478
221 129		**XC**	VL	*XC*	CZ	60379	60779	60979	60879	60479
221 130		**XC**	VL	*XC*	CZ	60380	60780	60980	60880	60480
221 131		**XC**	VL	*XC*	CZ	60381	60781	60981	60881	60481
221 132		**XC**	VL	*XC*	CZ	60382	60782	60982	60882	60482
221 133		**XC**	VL	*XC*	CZ	60383	60783	60983	60883	60483
221 134		**XC**	VL	*XC*	CZ	60384	60784	60984	60884	60484
221 135		**XC**	VL	*XC*	CZ	60385	60785	60985	60885	60485
221 136		**XC**	VL	*XC*	CZ	60386	60786	60986	60886	60486
221 137		**XC**	VL	*XC*	CZ	60387	60787	60987	60887	60487
221 138		**XC**	VL	*XC*	CZ	60388	60788	60988	60888	60488
221 139		**XC**	VL	*XC*	CZ	60389	60789	60989	60889	60489
221 140		**XC**	VL	*XC*	CZ	60390	60790	60990	60890	60490
221 141		**XC**	VL	*XC*	CZ	60391	60791	60991		60491
221 142	*	**VT**	VL	*VW*	CZ	60392	60992	60994	60792	60492
221 143	*	**VT**	VL	*VW*	CZ	60393	60993	60794	60793	60493
Spare	*	**VT**	VL		CZ	60394				60494

Names (carried on MS No. 609xx):

221 101	Louis Bleriot	221 109	Marco Polo
221 102	John Cabot	221 110	James Cook
221 103	Christopher Columbus	221 111	Roald Amundsen
221 104	Sir John Franklin	221 112	Ferdinand Magellan
221 105	William Baffin	221 113	Sir Walter Raleigh
221 106	Willem Barents	221 115	Polmadie Depot
221 107	Sir Martin Frobisher	221 142	BOMBARDIER Voyager
221 108	Sir Ernest Shackleton	221 143	Auguste Picard

CLASS 222   MERIDIAN   BOMBARDIER

Construction: Steel.
Engine: Cummins QSK19 of 560 kW (750 hp) at 1800 rpm.
Transmission: Two Alstom Onix 800 three-phase traction motors of 275 kW.
Braking: Rheostatic and electro-pneumatic.
Bogies: Bombardier B5005.   **Dimensions:** 23.85/23.00 x 2.73 m.
Couplers: Dellner at outer ends, bar within unit.
Gangways: Within unit only.   **Wheel Arrangement:** All cars 1A-A1.
Doors: Single-leaf swing plug.   **Maximum Speed:** 125 mph.
Seating Layout: 1: 2+1, 2: 2+2 facing/unidirectional.
Multiple Working: Within class and with Classes 220 and 221 (in an emergency).

222 001–006. 7-car units. DMF–MF–MF–MSRMB–MS–MS–DMS.

The 7-car units were built as 9-car units, before being reduced to 8-car sets and then later to 7-car sets to strengthen all 4-car units to 5-cars. 222 007 was built as a 9-car unit but later reduced to a 5-car unit.

DMRF. Bombardier Bruges 2004–05. 22/– 1TD 1W. 52.8 t.
MF. Bombardier Bruges 2004–05. 42/– 1T. 46.8 t.
MSRMB. Bombardier Bruges 2004–05. –/62. 48.0 t.
MS. Bombardier Bruges 2004–05. –/68 1T. 47.0 t.
DMS. Bombardier Bruges 2004–05. –/38 1TD 1W. 49.4 t.

222 001	**ST**	E	*EM*	DY	60241	60445	60341	60621
					60561	60551	60161	
222 002	**ST**	E	*EM*	DY	60242	60346	60342	60622
					60562	60544	60162	
222 003	**ST**	E	*EM*	DY	60243	60446	60343	60623
					60563	60553	60163	
222 004	**ST**	E	*EM*	DY	60244	60345	60344	60624
					60564	60554	60164	
222 005	**ST**	E	*EM*	DY	60245	60347	60443	60625
					60555	60565	60165	
222 006	**ST**	E	*EM*	DY	60246	60447	60441	60626
					60566	60556	60166	

Names (carried on MSRMB or DMS (222 003)):

222 001   THE ENTREPRENEUR EXPRESS
222 002   THE CUTLERS' COMPANY
222 003   TORNADO
222 004   CHILDREN'S HOSPITAL SHEFFIELD
222 006   THE CARBON CUTTER

222 007–023. 5-car units. DMF–MC–MSRMB–MS–DMS.

DMRF. Bombardier Bruges 2003–04. 22/– 1TD 1W. 52.8 t.
MC. Bombardier Bruges 2003–04. 28/22 1T. 48.6 t.
MSRMB. Bombardier Bruges 2003–04. –/62. 49.6 t.
MS. Bombardier Bruges 2004–05. –/68 1T. 47.0 t.
DMS. Bombardier Bruges 2003–04. –/40 1TD 1W. 51.0 t.

222 007	**ST**	E	*EM*	DY	60247	60442	60627	60567	60167
222 008	**ST**	E	*EM*	DY	60248	60918	60628	60545	60168
222 009	**ST**	E	*EM*	DY	60249	60919	60629	60557	60169
222 010	**ST**	E	*EM*	DY	60250	60920	60630	60546	60170
222 011	**ST**	E	*EM*	DY	60251	60921	60631	60531	60171
222 012	**ST**	E	*EM*	DY	60252	60922	60632	60532	60172
222 013	**ST**	E	*EM*	DY	60253	60923	60633	60533	60173
222 014	**ST**	E	*EM*	DY	60254	60924	60634	60534	60174
222 015	**ST**	E	*EM*	DY	60255	60925	60635	60535	60175
222 016	**ST**	E	*EM*	DY	60256	60926	60636	60536	60176
222 017	**ST**	E	*EM*	DY	60257	60927	60637	60537	60177
222 018	**ST**	E	*EM*	DY	60258	60928	60638	60444	60178
222 019	**ST**	E	*EM*	DY	60259	60929	60639	60547	60179
222 020	**ST**	E	*EM*	DY	60260	60930	60640	60543	60180

222 021	**ST**	E	*EM*	DY	60261 60931 60641 60552 60181
222 022	**ST**	E	*EM*	DY	60262 60932 60642 60542 60182
222 023	**ST**	E	*EM*	DY	60263 60933 60643 60541 60183

Names (carried on MSRMB or DMS (222 015)):

222 008 Derby Etches Park
222 015 175 YEARS OF DERBY'S RAILWAYS 1839–2014
222 022 INVEST IN NOTTINGHAM

222 101–104. 4-car former Hull Trains units. DMF–MC–MSRMB–DMS.

DMRF. Bombardier Bruges 2005. 22/– 1TD 1W. 52.8 t.
MC. Bombardier Bruges 2005. 11/46 1T. 47.1 t.
MSRMB. Bombardier Bruges 2005. –/62. 48.0 t.
DMS. Bombardier Bruges 2005. –/40 1TD 1W. 49.4 t.

222 101	**ST**	E	*EM*	DY	60271 60571 60681 60191
222 102	**ST**	E	*EM*	DY	60272 60572 60682 60192
222 103	**ST**	E	*EM*	DY	60273 60573 60683 60193
222 104	**ST**	E	*EM*	DY	60274 60574 60684 60194

3. SERVICE DMUS

This section lists vehicles not used for passenger-carrying purposes.
Vehicles are numbered in the special service stock number series.

CLASS 950   TRACK ASSESSMENT UNIT

DM–DM. Purpose built service unit based on the Class 150/1 design.
Gangwayed within unit.

Construction: Steel.
Engine: One Cummins NT-855-RT5 of 213 kW (285 hp) at 2100 rpm per
power car.
Transmission: Hydraulic. Voith T211r with cardan shafts to Gmeinder
GM190 final drive.
Maximum Speed: 75 mph.   **Couplers:** BSI automatic.
Bogies: BP38 (powered), BT38 (non-powered).
Brakes: Electro-pneumatic.   **Dimensions:** 20.06 x 2.82 m.
Doors: Manually operated slam & power operated sliding.
Multiple Working: Classes 142, 143, 144, 150, 153, 155, 156, 158, 159 and 170.

999600. DM. Lot No. 4060 BREL York 1987. 35.0 t.
999601. DM. Lot No. 4061 BREL York 1987. 35.0 t.

| 950 001 | **Y** | NR | *DB* | ZA | 999600 999601 |

CLASS 960   ROUTE LEARNING UNIT

DMB. Converted from Class 121. Non-gangwayed.

For details see Page 10.

This unit is available for hire to other operators for route learning if required.

Lot No. 30518 Pressed Steel 1960. 38.0 t.

960 014   **BG**   CR   *CR*   AL   977873   (55022)

CLASS 960   WATER-JETTING UNIT

DMB–MS–DMB. Converted 2003–04 from Class 117. Non-gangwayed.

Construction: Steel.
Engines: Two Leyland 1595 of 112 kW (150 hp) at 1800 rpm.
Transmission: Mechanical. Cardan shaft and freewheel to a four-speed epicyclic gearbox with a further cardan shaft to the final drive, each engine driving the inner axle of one bogie.
Maximum Speed: 70 mph.
Bogies: DD10.   **Couplers:** Screw.
Brakes: Twin pipe vacuum.   **Multiple Working:** Blue Square.
Doors: Manually operated slam.   **Dimensions:** 20.45 x 2.84 m.

977987/988. DMB. Lot No. 30546/30548 Pressed Steel 1959–60. 36.5 t.
977992. MS. Lot No. 30548 Pressed Steel 1959–60. 36.5 t.

960 301   **G**   CR   *CR*   AL   977987 (51371) 977992 (51375)
                                               977988 (51413)

4. DMUS AWAITING DISPOSAL

The list below comprises vehicles which are stored awaiting disposal.

Class 121

121 032   **AV**   CR   AL   55032

Class 960

Converted from Class 121.

960 011   **RK**   CR   TM   977859   (55025)

5. ON-TRACK MACHINES

These machines are used for maintaining, renewing and enhancing the infrastructure of the national railway network. With the exception of snowploughs all can be self-propelled, controlled either from a cab mounted on the machine or remotely. They are permitted to operate either under their own power or in train formations throughout the network both within and outside engineering possessions. Machines only permitted to be used within engineering possessions, referred to as On-Track Plant, are not included. Also not included are wagons included in OTM consists.

For each machine its Network Rail registered number, owner or responsible custodian and type is given, plus its name if carried. In addition, for snow clearance equipment and breakdown cranes the berthing location is given. Actual operation of each machine is undertaken by either the owner/responsible custodian or a contracted responsible custodian.

Machines were numbered by British Rail with either six-digit wagon series numbers or in the CEPS (Civil Engineers Plant System) series with five prefixed digits. Recently delivered machines have been numbered in the EVN series. Machines may also carry additional identifying numbers which are shown as "xxxx". Machines are listed here in CEPS/wagon series order. Those with EVN numbers are included where they would have been if allocated CEPS numbers.

(S) after the registered number designates a machine that is currently stored (the storage location of each is given at the end of this section).

DYNAMIC TRACK STABILISERS

DR 72211	BB	Plasser & Theurer DGS 62-N	
DR 72213	BB	Plasser & Theurer DGS 62-N	

TAMPERS

DR 73108	CS	Plasser & Theurer 09-32-RT	Tiger
DR 73109	SK	Plasser & Theurer 09-3X-RT	
DR 73110	SK	Plasser & Theurer 09-3X-RT	PETER WHITE Reading Panel 1965–2005
DR 73111	NR	Plasser & Theurer 09-3X-Dynamic	
DR 73113	NR	Plasser & Theurer 09-3X-Dynamic	
DR 73114	NR	Plasser & Theurer 09-3X-Dynamic	Ron Henderson
DR 73115	NR	Plasser & Theurer 09-3X-Dynamic	
DR 73116	NR	Plasser & Theurer 09-3X-Dynamic	
DR 73117	NR	Plasser & Theurer 09-3X Dynamic	
DR 73118	NR	Plasser & Theurer 09-3X Dynamic	
DR 73803	SK	Plasser & Theurer 08-32U-RT	Alexander Graham Bell
DR 73804	SK	Plasser & Theurer 08-32U-RT	James Watt
DR 73805	CS	Plasser & Theurer 08-16/32U-RT	
DR 73806	CS	Plasser & Theurer 08-16/32U-RT	Karine
DR 73904	SK	Plasser & Theurer 08-4x4/4S-RT	Thomas Telford
DR 73905	CS	Plasser & Theurer 08-4x4/4S-RT	Eddie King

DR 73906	CS	Plasser & Theurer 08-4x4/4S-RT	Panther
DR 73907	CS	Plasser & Theurer 08-4x4/4S-RT	
DR 73908	CS	Plasser & Theurer 08-4x4/4S-RT	
DR 73909	CS	Plasser & Theurer 08-4x4/4S-RT	Saturn
DR 73910	CS	Plasser & Theurer 08-4x4/4S-RT	Jupiter
DR 73911	CS	Plasser & Theurer 08-16/4x4C-RT	Puma
DR 73912	CS	Plasser & Theurer 08-16/4x4C-RT	Lynx
DR 73913	CS	Plasser & Theurer 08-12/4x4C-RT	
DR 73914	SK	Plasser & Theurer 08-4x4/4S-RT	Robert McAlpine
DR 73915	SK	Plasser & Theurer 08-16/4x4C-RT	William Arrol
DR 73916	SK	Plasser & Theurer 08-16/4x4C-RT	First Engineering
DR 73917	BB	Plasser & Theurer 08-4x4/4S-RT	
DR 73918	BB	Plasser & Theurer 08-4x4/4S-RT	
DR 73919	CS	Plasser & Theurer 08-16/4x4C100-RT	
DR 73920	CS	Plasser & Theurer 08-16/4x4C80-RT	
DR 73921	CS	Plasser & Theurer 08-16/4x4C80-RT	
DR 73922	CS	Plasser & Theurer 08-16/4x4C80-RT	John Snowdon
DR 73923	CS	Plasser & Theurer 08-4x4/4S-RT	Mercury
DR 73924	CS	Plasser & Theurer 08-16/4x4C100-RT	Atlas
DR 73925	CS	Plasser & Theurer 08-16/4x4C100-RT	Europa
DR 73926	BB	Plasser & Theurer 08-16/4x4C100-RT	Stephen Keith Blanchard
DR 73927	BB	Plasser & Theurer 08-16/4x4C100-RT	
DR 73928	BB	Plasser & Theurer 08-16/4x4C100-RT	
DR 73929	CS	Plasser & Theurer 08-4x4/4S-RT	
DR 73930	CS	Plasser & Theurer 08-4x4/4S-RT	
DR 73931	CS	Plasser & Theurer 08-16/4x4C100-RT	
DR 73932	SK	Plasser & Theurer 08-4x4/4S-RT	
DR 73933	SK	Plasser & Theurer 08-16/4x4/C100-RT	
DR 73934	SK	Plasser & Theurer 08-16/4x4/C100-RT	
DR 73935	CS	Plasser & Theurer 08-4x4/4S-RT	
DR 73936	CS	Plasser & Theurer 08-4x4/4S-RT	
DR 73937	BB	Plasser & Theurer 08-16/4x4C100-RT	
DR 73938	BB	Plasser & Theurer 08-16/4x4C100-RT	
DR 73939	BB	Plasser & Theurer 08-16/4x4C100-RT	Pat Best
DR 73940	SK	Plasser & Theurer 08-4x4/4S-RT	
DR 73941	SK	Plasser & Theurer 08-4x4/4S-RT	
DR 73942	CS	Plasser & Theurer 08-4x4/4S-RT	
DR 73943	BB	Plasser & Theurer 08-16/4x4C100-RT	
DR 73944	BB	Plasser & Theurer 08-16/4x4C100-RT	
DR 73945	BB	Plasser & Theurer 08-16/4x4C100-RT	
DR 73946	VO	Plasser & Theurer Euromat 08-4x4/4S	
DR 73947	CS	Plasser & Theurer 08-4x4/4S-RT	
DR 73948	CS	Plasser & Theurer 08-4x4/4S-RT	

99 70 9128 001-3 SK Plasser & Theurer Unimat 09-4x4/4S Dynamic "928001"

DR 75301	VO	Matisa B 45 UE
DR 75302	VO	Matisa B 45 UE
DR 75303	VO	Matisa B 45 UE
DR 75401	VO	Matisa B 41 UE
DR 75402	VO	Matisa B 41 UE
DR 75403	VO	Matisa B 41 UE

DR 75404	VO	Matisa B 41 UE
DR 75405	VO	Matisa B 41 UE
DR 75406	CS	Matisa B 41 UE
DR 75407	CS	Matisa B 41 UE
DR 75408	BB	Matisa B 41 UE
DR 75409	BB	Matisa B 41 UE
DR 75410	BB	Matisa B 41 UE
DR 75411	BB	Matisa B 41 UE
DR 75501	BB	Matisa B 66 UC
DR 75502	BB	Matisa B 66 UC

Eric Machell

BALLAST CLEANERS

DR 76323	NR	Plasser & Theurer RM95-RT
DR 76324	NR	Plasser & Theurer RM95-RT
DR 76501	NR	Plasser & Theurer RM-900-RT *(works with DR 92285/DR 92286)*
DR 76502	NR	Plasser & Theurer RM-900-RT *(works with DR 76702/DR 92331/ DR 92332)*
DR 76503	NR	Plasser & Theurer RM-900-RT *(works with DR 76701/DR 76801/ DR 92431/DR 92432)*

VACUUM PREPARATION MACHINES

DR 76701	NR	Plasser & Theurer VM80-NR *(works with DR 76503/DR 76801/ DR 92431/DR 92432)*
DR 76702	NR	Plasser & Theurer VM80-NR *(works with DR 76502/DR 92331/ DR 92332)*
DR 76703 (S)	NR	Plasser & Theurer VM80-NR
DR 76710 (S)	NR	Plasser & Theurer VM80-TRS
DR 76711 (S)	NR	Plasser & Theurer VM80-TRS

RAIL VACUUM MACHINES

| 99 70 9515 001-4 | BY | Railcare RA7-UK RailVac |
| 99 70 9515 002-2 | BY | Railcare RA7-UK RailVac |

BALLAST TRANSFER MACHINES

| DR 76750 | NR | Matisa D75 | *(works with DR 78802/DR 78812/ DR 78822/DR 78832)* |
| DR 76751 | NR | Matisa D75 | *(works with DR 78801/DR 78811/ DR 78821/DR 78831)* |

CONSOLIDATION MACHINE

| DR 76801 | NR | Plasser & Theurer 09-CM-NR | *(works with DR 76503/DR 76701/ DR 92431/DR 92432)* |

FINISHING MACHINES & BALLAST REGULATORS

DR 77001	SK	Plasser & Theurer AFM 2000-RT Finishing Machine
DR 77002	SK	Plasser & Theurer AFM 2000-RT Finishing Machine

DR 77315 (S)	BB	Plasser & Theurer USP 5000C Regulator
DR 77316 (S)	BB	Plasser & Theurer USP 5000C Regulator
DR 77322	BB	Plasser & Theurer USP 5000C Regulator
DR 77327	CS	Plasser & Theurer USP 5000C Regulator
DR 77336 (S)	BB	Plasser & Theurer USP 5000C Regulator
DR 77801	VO	Matisa R 24 S Regulator
DR 77802	VO	Matisa R 24 S Regulator
DR 77901	CS	Plasser & Theurer USP 5000-RT Regulator
DR 77903	NR	Plasser & Theurer USP 5000-RT Regulator Frank Jones
DR 77904	NR	Plasser & Theurer USP 5000-RT Regulator
DR 77905	NR	Plasser & Theurer USP 5000-RT Regulator
DR 77906	NR	Plasser & Theurer USP 5000-RT Regulator
DR 77907	NR	Plasser & Theurer USP 5000-RT Regulator
DR 77908	SK	Plasser & Theurer USP 5000-RT Regulator

TWIN JIB TRACK RELAYERS

DRP 78213	VO	Plasser & Theurer Self-Propelled Heavy Duty
DRP 78215	SK	Plasser & Theurer Self-Propelled Heavy Duty
DRP 78216	BB	Plasser & Theurer Self-Propelled Heavy Duty
DRP 78217 (S)	SK	Plasser & Theurer Self-Propelled Heavy Duty
DRP 78218 (S)	BB	Plasser & Theurer Self-Propelled Heavy Duty
DRP 78219	SK	Plasser & Theurer Self-Propelled Heavy Duty
DRP 78221	BB	Plasser & Theurer Self-Propelled Heavy Duty
DRP 78222	BB	Plasser & Theurer Self-Propelled Heavy Duty
DRP 78223 (S)	BB	Plasser & Theurer Self-Propelled Heavy Duty
DRP 78224 (S)	BB	Plasser & Theurer Self-Propelled Heavy Duty
DRC 78226	CS	Cowans Sheldon Self-Propelled Heavy Duty
DRC 78229	NR	Cowans Sheldon Self-Propelled Heavy Duty
DRC 78231	NR	Cowans Sheldon Self-Propelled Heavy Duty
DRC 78234	NR	Cowans Sheldon Self-Propelled Heavy Duty
DRC 78235	CS	Cowans Sheldon Self-Propelled Heavy Duty
DRC 78237 (S)	NR	Cowans Sheldon Self-Propelled Heavy Duty

NEW TRACK CONSTRUCTION
TRAIN PROPULSION MACHINES

DR 78701	BB	Harsco Track Technologies NTC-PW
DR 78702	BB	Harsco Track Technologies NTC-PW

TRACK RENEWAL MACHINES

Matisa P95 Track Renewals Trains
DR 78801+DR 78811+DR 78821+DR 78831   NR *(works with DR 76751)*
DR 78802+DR 78812+DR 78822+DR 78832   NR *(works with DR 76750)*

RAIL GRINDING TRAINS

Loram SPML 15
DR 79200A + DR 79200B + DR 79200C   NR

Loram SPML 17
DR 79201A + DR 79201B   NR

Speno RPS-32
DR 79221 + DR 79222 + DR 79223 + DR 79224 + DR 79225 + DR 79226   SI

Loram C21
DR 79231 + DR 79232 + DR 79233 + DR 79234 + DR 79235 + DR 79236 + DR 79237   NR
DR 79241 + DR 79242 + DR 79243 + DR 79244 + DR 79245 + DR 79246 + DR 79247   NR
DR 79251 + DR 79252 + DR 79253 + DR 79254 + DR 79255 + DR 79256 + DR 79257   NR

Names: DR 79241/247   Roger South *(one plate on opposite sides of each)*
          DR 79257   Martin Ellwood

Harsco Track Technologies RGH20C
DR 79261 + DR 79271   NR
DR 79262 + DR 79272   NR
DR 79263 + DR 79273   NR
DR 79264 + DR 79274   NR
DR 79265 (S)   NR   *spare vehicle*
DR 79267 + DR 79277   NR

STONEBLOWERS

DR 80200 (S)   NR   Pandrol Jackson Plain Line
DR 80201   NR   Pandrol Jackson Plain Line
DR 80202 (S)   NR   Pandrol Jackson Plain Line
DR 80203 (S)   NR   Pandrol Jackson Plain Line
DR 80204 (S)   NR   Pandrol Jackson Plain Line
DR 80205   NR   Pandrol Jackson Plain Line
DR 80206   NR   Pandrol Jackson Plain Line
DR 80207 (S)   NR   Pandrol Jackson Plain Line
DR 80208   NR   Pandrol Jackson Plain Line
DR 80209   NR   Pandrol Jackson Plain Line
DR 80210   NR   Pandrol Jackson Plain Line
DR 80211   NR   Pandrol Jackson Plain Line
DR 80212 (S)   NR   Pandrol Jackson Plain Line
DR 80213   NR   Harsco Track Technologies Plain Line
DR 80214   NR   Harsco Track Technologies Plain Line
DR 80215   NR   Harsco Track Technologies Plain Line
DR 80216   NR   Harsco Track Technologies Plain Line
DR 80217   NR   Harsco Track Technologies Plain Line

DR 80301	NR	Harsco Track Technologies Multi-purpose	Stephen Cornish
DR 80302	NR	Harsco Track Technologies Multi-purpose	
DR 80303	NR	Harsco Track Technologies Multi-purpose	

CRANES

DRP 81505	BB	Plasser & Theurer 12 tonne Heavy Duty Diesel Hydraulic
DRP 81507 (S)	BB	Plasser & Theurer 12 tonne Heavy Duty Diesel Hydraulic
DRP 81508 (S)	BB	Plasser & Theurer 12 tonne Heavy Duty Diesel Hydraulic
DRP 81511 (S)	BB	Plasser & Theurer 12 tonne Heavy Duty Diesel Hydraulic
DRP 81513 (S)	BB	Plasser & Theurer 12 tonne Heavy Duty Diesel Hydraulic
DRP 81517	BB	Plasser & Theurer 12 tonne Heavy Duty Diesel Hydraulic
DRP 81519 (S)	BB	Plasser & Theurer 12 tonne Heavy Duty Diesel Hydraulic
DRP 81522	BB	Plasser & Theurer 12 tonne Heavy Duty Diesel Hydraulic
DRP 81525	BB	Plasser & Theurer 12 tonne Heavy Duty Diesel Hydraulic
DRP 81532	BB	Plasser & Theurer 12 tonne Heavy Duty Diesel Hydraulic
DRK 81601	VO	Kirow KRC 810UK 100 tonne Heavy Duty Diesel Hydraulic
DRK 81602	BB	Kirow KRC 810UK 100 tonne Heavy Duty Diesel Hydraulic
DRK 81611	BB	Kirow KRC 1200UK 125 tonne Heavy Duty Diesel Hydraulic
DRK 81612	CS	Kirow KRC 1200UK 125 tonne Heavy Duty Diesel Hydraulic
DRK 81613	VO	Kirow KRC 1200UK 125 tonne Heavy Duty Diesel Hydraulic
DRK 81621	VO	Kirow KRC 250UK 25 tonne Diesel Hydraulic
DRK 81622	VO	Kirow KRC 250UK 25 tonne Diesel Hydraulic
DRK 81623	SK	Kirow KRC 250UK 25 tonne Diesel Hydraulic
DRK 81624	SK	Kirow KRC 250UK 25 tonne Diesel Hydraulic
DRK 81625	SK	Kirow KRC 250UK 25 tonne Diesel Hydraulic

Names:

| DRK 81601 | Nigel Chester | | DRK 81611 | Malcolm L. Pearce |

LONG WELDED RAIL TRAIN PROPULSION MACHINES

DR 89005	NR	Cowans Boyd PW
DR 89006 (S)	NR	Cowans Boyd PW
DR 89007	NR	Cowans Boyd PW
DR 89008	NR	Cowans Boyd PW
DR 89009 (S)	NR	Cowans Boyd PW

BALLAST SYSTEM PROPULSION MACHINES

DR 92263	NR	Plasser & Theurer MFS-PW
DR 92264	NR	Plasser & Theurer NB-PW
DR 92285	NR	Plasser & Theurer PW-RT *(works with DR 76501/DR 92286)*
DR 92286	NR	Plasser & Theurer NPW-RT *(works with DR 76501/DR 92285)*
DR 92331	NR	Plasser & Theurer PW-RT *(works with DR 76502/DR 76702/ DR 92332)*
DR 92332	NR	Plasser & Theurer NPW-RT *(works with DR 76502/DR 76702/ DR 92331)*
DR 92431	NR	Plasser & Theurer PW-RT *(works with DR 76503/DR 76701/ DR 76801/DR 92432)*
DR 92432	NR	Plasser & Theurer NPW-RT *(works with DR 76503/DR 76701/ DR 76801/DR 92431)*

TELESCOPIC BREAKDOWN CRANES

ADRC 96710	NR	Cowans Sheldon 75 tonne Diesel Hydraulic	Wigan Springs Branch
ADRC 96713	NR	Cowans Sheldon 75 tonne Diesel Hydraulic	Wigan Springs Branch
ADRC 96714	NR	Cowans Sheldon 75 tonne Diesel Hydraulic	Knottingley Depot
ADRC 96715	NR	Cowans Sheldon 75 tonne Diesel Hydraulic	Bescot Depot

GENERAL PURPOSE VEHICLES

DR 97001	H1	Eiv de Brieve DU94BA TRAMM with Crane "DU 94 B 001 URS"
DR 97011	H1	Windhoff MPV (Modular)
DR 97012	H1	Windhoff MPV (Modular)
DR 97013	H1	Windhoff MPV (Modular)
DR 97014	H1	Windhoff MPV (Modular)
DR 98008	NR	Windhoff MPV Twin-cab with test equipment
DR 98215A + DR 98215B	BB	Plasser & Theurer GP-TRAMM with Trailer
DR 98216A + DR 98216B	BB	Plasser & Theurer GP-TRAMM with Trailer
DR 98217A + DR 98217B	BB	Plasser & Theurer GP-TRAMM with Trailer
DR 98218A + DR 98218B	BB	Plasser & Theurer GP-TRAMM with Trailer
DR 98219A + DR 98219B	BB	Plasser & Theurer GP-TRAMM with Trailer
DR 98220A + DR 98220B	BB	Plasser & Theurer GP-TRAMM with Trailer
DR 98305 (S) NR		Geismar GP-TRAMM VMT 860 PL/UM
DR 98306 (S) NR		Geismar GP-TRAMM VMT 860 PL/UM
DR 98307A + DR 98307B (S) CS		Geismar GP-TRAMM VMT 860 PL/UM with Trailer
DR 98308A + DR 98308B (S) CS		Geismar GP-TRAMM VMT 860 PL/UM with Trailer
DR 98901 + DR 98951	NR	Windhoff MPV Master & Slave
DR 98902 + DR 98952	NR	Windhoff MPV Master & Slave
DR 98903 + DR 98953	NR	Windhoff MPV Master & Slave
DR 98904 + DR 98954	NR	Windhoff MPV Master & Slave
DR 98905 + DR 98955	NR	Windhoff MPV Master & Slave
DR 98906 + DR 98956	NR	Windhoff MPV Master & Slave
DR 98907 + DR 98957	NR	Windhoff MPV Master & Slave
DR 98908 + DR 98958	NR	Windhoff MPV Master & Slave
DR 98909 + DR 98959	NR	Windhoff MPV Master & Slave
DR 98910 + DR 98960	NR	Windhoff MPV Master & Slave
DR 98911 + DR 98961	NR	Windhoff MPV Master & Slave
DR 98912 + DR 98962	NR	Windhoff MPV Master & Slave
DR 98913 + DR 98963	NR	Windhoff MPV Master & Slave
DR 98914 + DR 98964	NR	Windhoff MPV Master & Slave
DR 98915 + DR 98965	NR	Windhoff MPV Master & Slave
DR 98916 + DR 98966	NR	Windhoff MPV Master & Slave
DR 98917 + DR 98967	NR	Windhoff MPV Master & Slave
DR 98918 + DR 98968	NR	Windhoff MPV Master & Slave
DR 98919 + DR 98969	NR	Windhoff MPV Master & Slave
DR 98920 + DR 98970	NR	Windhoff MPV Master & Slave
DR 98921 + DR 98971	NR	Windhoff MPV Master & Slave
DR 98922 + DR 98972	NR	Windhoff MPV Master & Slave

DR 98923 + DR 98973	NR	Windhoff MPV Master & Slave
DR 98924 + DR 98974	NR	Windhoff MPV Master & Slave
DR 98925 + DR 98975	NR	Windhoff MPV Master & Slave
DR 98926 + DR 98976	NR	Windhoff MPV Master & Powered Slave
DR 98927 + DR 98977	NR	Windhoff MPV Master & Powered Slave
DR 98928 + DR 98978	NR	Windhoff MPV Master & Powered Slave
DR 98929 + DR 98979	NR	Windhoff MPV Master & Powered Slave
DR 98930 + DR 98980	NR	Windhoff MPV Master & Powered Slave
DR 98931 + DR 98981	NR	Windhoff MPV Master & Powered Slave
DR 98932 + DR 98982	NR	Windhoff MPV Master & Powered Slave

Name:

DR 97012   Geoff Bell

ELECTRIFICATION VEHICLES

DR 98001	NR	Windhoff MPV with Piling Equipment
DR 98002 (S)	NR	Windhoff MPV with Piling Equipment
DR 98003	NR	Windhoff MPV with Overhead Line Renewal Equipment
DR 98004	NR	Windhoff MPV with Overhead Line Renewal Equipment
DR 98005 (S)	NR	Windhoff MPV with Piling Equipment
DR 98006	NR	Windhoff MPV with Piling Equipment
DR 98007	NR	Windhoff MPV with Piling Equipment
DR 98009	NR	Windhoff MPV with Overhead Line Renewal Equipment
DR 98010	NR	Windhoff MPV with Overhead Line Renewal Equipment
DR 98011	NR	Windhoff MPV with Overhead Line Renewal Equipment
DR 98012	NR	Windhoff MPV with Overhead Line Renewal Equipment
DR 98013	NR	Windhoff MPV with Overhead Line Renewal Equipment
DR 98014	NR	Windhoff MPV with Overhead Line Renewal Equipment

99 70 9131 001-8	NR	Windhoff MPV with Piling Equipment	"DR 76901"
99 70 9131 003-4	NR	Windhoff MPV with Piling Equipment	"DR 76903"
99 70 9131 005-9	NR	Windhoff MPV with Piling Equipment	"DR 76905"
99 70 9131 006-7	NR	Windhoff MPV with Concrete Equipment	"DR 76906"
99 70 9131 010-9	NR	Windhoff MPV with Concrete Equipment	"DR 76910"
99 70 9131 011-7	NR	Windhoff MPV with Structure Equipment	"DR 76911"
99 70 9131 013-3	NR	Windhoff MPV with Structure Equipment	"DR 76913"
99 70 9131 014-1	NR	Windhoff MPV with Overhead Line Equipment	"DR 76914"
99 70 9131 015-8	NR	Windhoff MPV with Overhead Line Equipment	"DR 76915"
99 70 9131 018-2	NR	Windhoff MPV with Overhead Line Equipment	"DR 76918"
99 70 9131 020-8	NR	Windhoff MPV with Overhead Line Equipment	"DR 76920"
99 70 9131 021-6	NR	Windhoff MPV with Overhead Line Equipment	"DR 76921"
99 70 9131 022-4	NR	Windhoff MPV with Final Works Equipment	"DR 76922"
99 70 9131 023-2	NR	Windhoff MPV with Final Works Equipment	"DR 76923"

Names:

DR 98926+DR 98976	John Denyer
DR 98003	ANTHONY WRIGHTON 1944–2011
DR 98004	PHILIP CATTRELL 1961–2011
DR 98009	MELVYN SMITH 1953–2011
DR 98010	BENJAMIN GAUTREY 1992–2011
99 70 9131 001-8	BRUNEL

SNOWPLOUGHS

ADB 965203	NR	Independent Drift Plough	Tees Yard
ADB 965206	NR	Independent Drift Plough	York Parcels Sidings
ADB 965208	NR	Independent Drift Plough	Motherwell Depot
ADB 965209	NR	Independent Drift Plough	Bristol Barton Hill Depot
ADB 965210	NR	Independent Drift Plough	Tonbridge West Yard
ADB 965211	NR	Independent Drift Plough	March Depot
ADB 965217	NR	Independent Drift Plough	Edinburgh Slateford Depot
ADB 965219	NR	Independent Drift Plough	Edinburgh Slateford Depot
ADB 965223	NR	Independent Drift Plough	Margam Wagon Works
ADB 965224	NR	Independent Drift Plough	Carlisle Kingmoor Depot
ADB 965230	NR	Independent Drift Plough	Carlisle Kingmoor Depot
ADB 965231	NR	Independent Drift Plough	Bristol Barton Hill Depot
ADB 965234	NR	Independent Drift Plough	Inverness Millburn Yard
ADB 965235	NR	Independent Drift Plough	Margam Wagon Works
ADB 965236	NR	Independent Drift Plough	Tonbridge West Yard
ADB 965237	NR	Independent Drift Plough	March Depot
ADB 965240	NR	Independent Drift Plough	Motherwell Depot
ADB 965241	NR	Independent Drift Plough	York Turntable Sidings
ADB 965242	NR	Independent Drift Plough	Tees Yard
ADB 965243	NR	Independent Drift Plough	Inverness Millburn Yard
ADB 965576	NR	Beilhack Type PB600 Plough	Doncaster West Yard
ADB 965577	NR	Beilhack Type PB600 Plough	Doncaster West Yard
ADB 965578	NR	Beilhack Type PB600 Plough	Carlisle Kingmoor Yard
ADB 965579	NR	Beilhack Type PB600 Plough	Carlisle Kingmoor Yard
ADB 965580	NR	Beilhack Type PB600 Plough	Crewe Gresty Bridge Depot
ADB 965581	NR	Beilhack Type PB600 Plough	Crewe Gresty Bridge Depot
ADB 966098	NR	Beilhack Type PB600 Plough	Doncaster West Yard
ADB 966099	NR	Beilhack Type PB600 Plough	Doncaster West Yard

SNOWBLOWERS

ADB 968500	NR	Beilhack Self-Propelled Rotary	Edinburgh Slateford Depot
ADB 968501	NR	Beilhack Self-Propelled Rotary	Edinburgh Slateford Depot

INFRASTRUCTURE MONITORING VEHICLES

999800 (S)	NR	Plasser & Theurer EM-SAT 100/RT Track Survey Car
999801 (S)	NR	Plasser & Theurer EM-SAT 100/RT Track Survey Car

Name: 999800   Richard Spoors

ON-TRACK MACHINES AWAITING DISPOSAL

Tampers

DR 73105	Plasser & Theurer 09-32 CSM	Rugby Depot
DR 73503	Plasser & Theurer 08-16/90 ZW	Ashford OTM Depot
DR 75201	Plasser & Theurer 08-275 S&C	Hither Green Depot
DR 75202	Plasser & Theurer 08-275 S&C	Hither Green Depot

Ballast Cleaners

DR 76304	Plasser & Theurer RM74	Plasser UK, West Ealing
DR 76318	Plasser & Theurer RM74	Plasser UK, West Ealing

Twin Jib track relayer

DRB 78123	British Hoist & Crane Non-Self-Propelled	Polmadie DHS

LOCATIONS OF STORED ON-TRACK MACHINES

The locations of machines shown above as stored (S) are shown here.

DR 76703	Toton Yard	DRP 81507	Ashford OTM Depot
DR 76710	Crewe Gresty Lane Sidings	DRP 81508	Ashford OTM Depot
DR 76711	Taunton Fairwater Yard	DRP 81511	Ashford OTM Depot
DR 77315	Ashford OTM Depot	DRP 81513	Ashford OTM Depot
DR 77316	Ashford OTM Depot	DRP 81519	Woking OTM Depot
DR 77336	Hither Green Depot	DR 89006	York Klondyke Yard
DRP 78217	Glasgow Rutherglen Depot	DR 89009	York Klondyke Yard
DRP 78218	Ashford OTM Depot	DR 98002	York Holgate Works
DRP 78223	Ashford OTM Depot	DR 98005	York Holgate Works
DRP 78224	Hither Green Depot	DR 98305	Eastleigh Works
DRC 78237	York Holgate Works	DR 98306	Eastleigh Works
DR 79265	Eastleigh Works	DR 98307A+	
DR 80200	East Dereham	DR 98307B	Rugby Depot
DR 80202	Eastleigh Works	DR 98308A+	
DR 80203	Eastleigh Works	DR 98308B	Rugby Depot
DR 80204	East Dereham	999800	Eastleigh Works
DR 80207	Eastleigh Works	999801	Eastleigh Works
DR 80212	Eastleigh Works		

6. UK LIGHT RAIL & METRO SYSTEMS

This section lists the rolling stock of the various light rail and metro systems in Great Britain. Passenger carrying vehicles only are covered (not works vehicles). This listing does not cover the London Underground network.

6.1. BLACKPOOL & FLEETWOOD TRAMWAY

Until the opening of Manchester Metrolink, the Blackpool tramway was the only urban/inter-urban tramway system left in Britain. The infrastructure is owned by Blackpool Corporation, and the tramway is operated by Blackpool Transport Services Ltd. The 11 miles from Fleetwood to Starr Gate reopened in spring 2012 following rebuilding as a modern light rail system with a new fleet of 16 Bombardier Flexity 2 trams: these are now used on all scheduled services.

System: 600 V DC overhead.
Depot & Workshops: Starr Gate and Rigby Road (heritage fleet).
Standard livery: Flexity 2s and **F:** White, black & purple.

FLEXITY 2   5-SECTION TRAMS

These 16 articulated Supertrams are the trams used in normal daily service.

Built: 2011–12 by Bombardier Transportation, Bautzen, Germany.
Wheel arrangement: Bo-2-Bo.
Traction Motors: 4 x Bombardier 3-phase asynchronous of 120 kW.
Dimensions: 32.2 x 2.65 m.   **Seats:** 70 (4).
Doors: Sliding plug.   **Couplers:**
Weight: 40.9 t.   **Maximum Speed:** 43 mph.
Braking: Regenerative, disc and magnetic track.

Advertising livery: 016 Fleetwood Freeport (blue).

001	005	009	013
002	006	010	014
003	007	011	015
004	008	012	016 **AL**

Name: 002   "Alderman E.E. Wynne"

▲ Arriva Trains-liveried 175 007 arrives at Stockport with the 14.30 Manchester Piccadilly–Milford Haven on 11/03/14.   **Robert Pritchard**

▼ First Great Western Dynamic Lines-liveried 180 102 passes Didcot North Junction with a London Paddington–Oxford train on 01/08/13.   **Andrew Mason**

▲ TransPennine Express 185 126 approaches Rotherham Masborough on 23/07/14 with the 17.26 Cleethorpes–Manchester Airport. **Robert Pritchard**

▼ Hastings Diesels preserved Class 201 DEMU 1001 rounds the Queensville Curve into Stafford with a 17.16 Crewe–Hastings railtour on 12/07/14. **Chris Morrison**

▲ CrossCountry 220 003 leaves Stafford on 20/08/13 with the 13.07 Manchester Piccadilly–Bristol Temple Meads. **Cliff Beeton**

▼ 221 126 passes Tyseley with the 14.40 Reading–Newcastle on 29/09/13.
Robert Pritchard

▲ East Midlands Trains-liveried 222 016 leaves Sheffield with the 16.49 to London St Pancras on 23/07/14. **Robert Pritchard**

▼ Colas Rail Plasser & Theurer 08-4x4/4S-RT Tamper DR 73936 passes Exeter St Thomas running from Kings Norton to Tavistock Junction on 14/02/14. **David Hunt**

▲ Volker Rail Matisa R 24 S Ballast Regulator DR 77802 passes Normanton-on-Soar, near Loughborough, working from Trent Sidings to Doncaster on 27/03/14. **Paul Biggs**

▼ Network Rail Loram SPML 17 Rail Grinding Train DR 79201 at Carlisle Upperby on 11/07/14. **Craig Millar**

▲ Network Rail Pandrol Jackson Plain Line Stoneblower DR 80201 heads north at Normanton working from Huntingdon to Lichfield City on 11/04/14. **Paul Biggs**

▼ The structural equipment section of the new Network Rail Electrification Train formation consisting of 99 70 9131 013-3 (leading) and 99 70 9131 011-7 (tailing) that also carry identities DR 76913 and DR 76911 respectively, are seen just south of Abbotswood Junction running from Tuxford to Swindon on 06/08/14.
Steve Widdowson

▲ Network Rail Independent Drift Snowploughs ADB 965243 and ADB 965234 are seen either end of DRS 37218 and 37606 at Feabuie on their way from Culloden to Inverness on 25/02/14. **Alexander Colley**

▼ Manchester Metrolink Bombardier Flexity Swift tram 3001 leaves East Didsbury with a service to Rochdale via Oldham on 25/05/13. **Robert Pritchard**

▲ One of the new Alstom Citadis Nottingham trams, 221, approaches High School stop with a Station Street service on 03/08/14.　　**Robert Pritchard**

▼ The first of the 20 new Midland Metro CAF trams entered service in September 2014. On 05/09/14 23 arrives at The Hawthorns for Birmingham Snow Hill.
　　Robert Pritchard

"BALLOON" DOUBLE DECKERS   A1-1A

The nine cars listed have partial exemption from the Rail Vehicle Accessibility Regulations and have been fitted with wider doors. However in 2014 all were stored, except 700 and 711 (which was loaned to the National Tramway Museum at Crich during 2014).

Built: 1934–35 by English Electric.
Traction Motors: 2 x EE305 of 40 kW. **Seats:** 94 (*† 92).

* Rebuilt with a flat front end design and air-conditioned cabs.

Advertising liveries:

707: Coral Island: The Jewel on the Mile (black)	718: Madame Tussaud's (purple & red)
709: Blackpool Sealife Centre (blue)	720: Walls ice cream (red)
713: Houndshill Shopping Centre (purple & white)	724: Lyndene Hotel (blue)

700	**F**		711 †	**F**		719	**F**	(S)
707 *	**AL** (S)		713	**AL** (S)		720	**AL**	(S)
709 *	**AL** (S)		718 *	**AL** (S)		724 *	**AL**	(S)

Name: 719   Donna's Dream House

HERITAGE FLEET of VINTAGE CARS

The following trams have exemption from the RVAR for Heritage use. They normally see use during the autumn Illuminations season or for private excursions.

Blackpool & Fleetwood 40	Single-deck Fleetwood Box car	Built: 1914
Bolton 66	Bogie double-decker	Built: 1901
Blackpool 147 MICHAEL AIREY	Standard double-decker	Built: 1924
Blackpool 227 (602)	Open boat car	Built: 1934
Blackpool 230 (604) GEORGE FORMBY OBE	Open boat car	Built: 1934
Blackpool 272+T2 (672+682)	Progress Twin Car	Rebuilt: 1960
Blackpool 304	Coronation Class single-decker	Built: 1952
Blackpool 600 THE DUCHESS OF CORNWALL	Open boat car	Built: 1934
Blackpool 631	Brush car	Built: 1937
Blackpool 642	Centenary car	Built: 1986
Blackpool 648	Centenary car	Built: 1987
Blackpool 701	Balloon double-decker	Built: 1934
Blackpool 706 PRINCESS ALICE	Balloon open-top double-decker	Built: 1934
Blackpool 717 WALTER LUFF	Balloon double-decker	Built: 1935
Blackpool 723	Balloon double-decker	Built: 1935

Illuminated cars

Blackpool 733	Western Train loco & tender	Rebuilt: 1962
Blackpool 734	Western Train coach	Rebuilt: 1962
Blackpool 736	"Warship" HMS Blackpool	Rebuilt: 1965
Blackpool 737	Illuminated Trawler – "Fisherman's Friend"	Rebuilt: 2001

STORED VEHICLES

The following vehicles are stored at Rigby Road depot.

Blackpool 8 (S)	BCT One Man Car	Rebuilt: 1974
Blackpool 143 (S)	Standard double-decker	Built: 1924
Blackpool 259 (S)	Brush car	Built: 1937
Blackpool 279 (S)	English Electric Railcoach	Rebuilt: 1960
Blackpool 622 (S)	Brush car	Built: 1937
Blackpool 632 (S)	Brush car	Built: 1937
Blackpool 660 (S)	Coronation Class single-decker	Built: 1953
Blackpool 663 (S)	Coronation Class single-decker	Built: 1953
Blackpool 675+685 (S)	Progress Twin Car	Rebuilt: 1958/60
Blackpool 676+686 (S)	Progress Twin Car	Rebuilt: 1958/60
Blackpool 704 (S)	Balloon double-decker	Built: 1934
Blackpool 715 (S)	Balloon double-decker	Built: 1935
Blackpool 732 (S)	Rocket illuminated car	Built: 1961
Blackpool 761 (S)	Jubilee Class double-decker	Rebuilt: 1979

6.2. DOCKLANDS LIGHT RAILWAY

This system runs for a total of approximately 23 route miles from termini at Bank and Tower Gateway in central London to Lewisham, Stratford, Beckton and Woolwich Arsenal. A new line from Canning Town to Stratford International also opened in 2011. The first part of the network opened in 1987 from Tower Gateway to Island Gardens. Originally owned by London Transport, it is now part of the London Rail division of Transport for London and currently operated by Serco (but will be operated by Keolis/Amey from December 2014). Cars are normally driven automatically using the Alcatel Seltrack moving block signalling system.

Original P86 and P89 Class vehicles 01–21 were withdrawn from service in 1991 (01–11) and 1995 (12–21) and sold for use in Essen, Germany. 55 new cars from Bombardier in Germany entered traffic between 2008 and 2010. These new vehicles enabled 3-unit trains to operate on all routes.

System: 750 V DC third rail (bottom contact). High-floor.
Depots: Beckton (main depot) and Poplar.
Livery: Red with a curving blue stripe to represent the River Thames.

CLASS B90   2-SECTION UNITS

Built: 1991–92 by BN Construction, Bruges, Belgium. Chopper control.
Wheel Arrangement: B-2-B.   **Traction Motors:** 2 x Brush 140 kW.
Seats: 52 (4).   **Weight:** 37t.
Dimensions: 28.80 x 2.65 m.   **Braking:** Rheostatic.
Couplers: Scharfenberg.   **Maximum Speed:** 50 mph.
Doors: Sliding. End doors for staff use.

22	26	30	34	38	42
23	27	31	35	39	43
24	28	32	36	40	44
25	29	33	37	41	

CLASS B92   2-SECTION UNITS

Built: 1992–95 by BN Construction, Bruges, Belgium. Chopper control.
Wheel Arrangement: B-2-B.   **Traction Motors:** 2 x Brush 140 kW.
Seats: 52 (4).   **Weight:** 37 t.
Dimensions: 28.80 x 2.65 m.   **Braking:** Rheostatic.
Couplers: Scharfenberg.   **Maximum Speed:** 50 mph.
Doors: Sliding. End doors for staff use.

45	53	61	69	77	85
46	54	62	70	78	86
47	55	63	71	79	87
48	56	64	72	80	88
49	57	65	73	81	89
50	58	66	74	82	90
51	59	67	75	83	91
52	60	68	76	84	

CLASS B2K   2-SECTION UNITS

Built: 2002–03 by Bombardier Transportation, Bruges, Belgium.
Wheel Arrangement: B-2-B.   **Traction Motors:** 2 x Brush 140 kW.
Seats: 52 (4).   **Weight:** 37 t.
Dimensions: 28.80 x 2.65 m.   **Braking:** Rheostatic.
Couplers: Scharfenberg.   **Maximum Speed:** 50 mph.
Doors: Sliding. End doors for staff use.

92	96	01	05	09	13
93	97	02	06	10	14
94	98	03	07	11	15
95	99	04	08	12	16

CLASS B07   2-SECTION UNITS

Built: 2007–10 by Bombardier Transportation, Bautzen, Germany.
Wheel Arrangement: B-2-B.   **Traction Motors:** 2 x Brush 140 kW.
Seats: 52 (4).   **Weight:** 37 t.
Dimensions: **Braking:** Rheostatic.
Couplers: Scharfenberg.   **Maximum Speed:** 50 mph.
Doors: Sliding. End doors for staff use.

101	111	120	129	138	147
102	112	121	130	139	148
103	113	122	131	140	149
104	114	123	132	141	150
105	115	124	133	142	151
106	116	125	134	143	152
107	117	126	135	144	153
108	118	127	136	145	154
109	119	128	137	146	155
110					

6.3. EDINBURGH TRAMWAY

The new Edinburgh tramway finally opened in May 2014, after years of delays. The scheme was dogged by construction problems, with the originally planned terminus of Newhaven in the north of the city later cut back to York Place in the city centre. In the west side of the city trams operate to Edinburgh Airport, the complete route being 8½ miles long.

The tramway is operated by Edinburgh Trams Ltd, a publicly owned company that works in partnership with Lothian Buses, as part of the Transport for Edinburgh Group.

The trams are the longest to operate in the UK, although only around 9 of the 27 are normally required to operate the service (but all are used as cars are rotated in traffic).

System: 750 V DC overhead.
Platform Height: 350 mm.
Depot & Workshops: Gogar.
Livery: White, red & black.

CAF   7-SECTION TRAMS

Built: 2009–11 by CAF, Irun, Spain.
Wheel Arrangement: Bo-Bo-2-Bo.   **Traction Motors:** 12 x CAF 80 kW.
Seats: 78.   **Weight:** 56.25 t.
Dimensions: 42.8 x 2.65 m.   **Braking:** Regenerative & electro hydraulic.
Couplers: Albert.   **Maximum Speed:** 50 mph.
Doors: Sliding plug.

251	257	263	268	273
252	258	264	269	274
253	259	265	270	275
254	260	266	271	276
255	261	267	272	277
256	262			

6.4. GLASGOW SUBWAY

This circular 4 ft gauge underground line is the smallest metro system in the UK, running for just over six miles. Operated by Strathclyde PTE the system has 15 stations. The entire passenger railway is underground, contained in twin tunnels, allowing for clockwise operation on the "outer" circle and anti-clockwise operation on the "inner" circle.

Trains are formed of 3-cars – either three power cars or two power cars sandwiching one of the newer trailer cars.

System: 600 V DC third rail.
Depot & Workshops: Broomloan.
Livery: Orange & grey Subway livery.

SINGLE POWER CARS

Built: 1977–79 by Metro-Cammell, Birmingham. Refurbished 1993–95 by ABB Derby.
Wheel Arrangement: Bo-Bo.
Traction Motors: 4 x GEC G312AZ of 35.6 kW.
Seats: 36.
Couplers: Wedglock.
Weight: 19.6 t.
Dimensions: 12.81 m x 2.34 m.
Doors: Sliding.
Maximum Speed: 33.5 mph.

101	108	115	122	128
102	109	116	123	129
103	110	117	124	130
104	111	118	125	131
105	112	119	126	132
106	113	120	127	133
107	114	121		

INTERMEDIATE BOGIE TRAILERS

Built: 1992 by Hunslet Barclay, Kilmarnock.
Seats: 40.
Couplers: Wedglock.
Weight: 17.2 t.
Dimensions: 12.70 m x 2.34 m.
Doors: Sliding.
Maximum Speed: 33.5 mph.

201	203	205	207	208
202	204	206		

6.5. GREATER MANCHESTER METROLINK

Metrolink was the first modern tramway system in the UK, combining street running with longer distance running over former BR lines. The system opened in 1992 from Bury to Altrincham with a street section through the centre of Manchester and a spur to Piccadilly station. A second line opened in 2000 from Cornbrook to Eccles.

A short spur off the Eccles line to MediaCityUK opened in September 2010 whilst the first part of the South Manchester Line to Chorlton and St Werburgh's Road opened in July 2011. This was followed by a further extension to East Didsbury in May 2013. In June 2012 the former National Rail line from Manchester to Oldham Mumps opened as a Metrolink line and this was extended to Shaw & Crompton in December 2012 and to Rochdale station in February 2013. The East Manchester Line reached Droylsden in February 2013 and this was followed by Droylsden–Ashton-under-Lyne in October 2013, Oldham town centre (January 2014) and Rochdale town centre (March 2014), extending the total route mileage to 48½ miles.

The next extension to open will be from St Werburgh's Road to Manchester Airport in late 2014. This will be followed by a second city crossing in Manchester, on which work is now underway.

Operator: RATP Dev.
System: 750 V DC overhead. High floor.
Depot & Workshops: Queens Road and Trafford.

T68 1000 SERIES   2-SECTION TRAMS

All T68 trams have now been withdrawn and those remaining in Manchester are all awaiting disposal.

Built: 1991–92 by Firema, Italy. Chopper control.

Wheel Arrangement: Bo-2-Bo.   **Traction Motors:** 4 x GEC 130 kW.
Dimensions: 29.0 x 2.65 m.   **Seats:** 82 (4).
Doors: Sliding.   **Couplers:** Scharfenberg.
Weight: 45 t.   **Maximum Speed:** 50 mph.
Braking: Regenerative, disc and emergency track.

Liveries: White, dark grey & blue with light blue doors.
M: Silver & yellow.

(S) Stored at Trafford depot (except 1003 at Queens Road).

1003 is reserved for the Greater Manchester Fire & Rescue Service.

1007 is reserved for Heaton Park Tramway.

1016, 1022, 1024 and 1026 have been moved to Long Marston for use in UKTram development work.

1003	**M** (S)	
1007	(S)	EAST LANCASHIRE RAILWAY
1017	(S)	
1020	(S)	
1021	(S)	
1023	(S)	

T68 2000 SERIES   2-SECTION TRAMS

Built: 1999 by Ansaldo, Italy. Chopper control.
Wheel Arrangement: Bo-2-Bo.   **Traction Motors:** 4 x GEC 130 kW.
Dimensions: 29.0 x 2.65 m.   **Seats:** 82 (4).
Doors: Sliding.   **Couplers:** Scharfenberg.
Weight: 45 t.   **Maximum Speed:** 50 mph.
Braking: Regenerative, disc and magnetic track.

Livery: White, dark grey & blue with light blue doors.

2001	(S)		2005	(S)
2002	(S)		2006	(S)
2003	(S) DAVE HANSFORD			

3000 SERIES   FLEXITY SWIFT   2-SECTION TRAMS

120 Bombardier M5000 "Flexity Swift" trams are currently being delivered. These trams now operate all services, having replaced the T68 series trams. The operate either singly or in pairs. Deliveries had reached 3088 by October 2014 and all 120 will be delivered by 2017. Trams from 3075 upwards have 8 more seats.

Built: 2009–16 by Bombardier, Vienna, Austria.
Wheel Arrangement: Bo-2-Bo.
Traction Motors: 4 x Bombardier 3-phase asynchronous of 120 kW.
Dimensions: 28.4 x 2.65 m.   **Seats:** 52 or 60 (3075–3120).
Doors: Sliding.   **Couplers:** Scharfenberg.
Weight: 39.7 t.   **Maximum Speed:** 50 mph.
Braking: Regenerative, disc and magnetic track.

Livery: Silver & yellow.

3001–3004 have been fitted with special "ice-breaking" pantographs.

3001	3025	3049	3073	3097
3002	3026	3050	3074	3098
3003	3027	3051	3075	3099
3004	3028	3052	3076	3100
3005	3029	3053	3077	3101
3006	3030	3054	3078	3102
3007	3031	3055	3079	3103
3008	3032	3056	3080	3104
3009	3033	3057	3081	3105
3010	3034	3058	3082	3106
3011	3035	3059	3083	3107
3012	3036	3060	3084	3108
3013	3037	3061	3085	3109
3014	3038	3062	3086	3110
3015	3039	3063	3087	3111
3016	3040	3064	3088	3112
3017	3041	3065	3089	3113
3018	3042	3066	3090	3114
3019	3043	3067	3091	3115
3020	3044	3068	3092	3116
3021	3045	3069	3093	3117
3022	3046	3070	3094	3118
3023	3047	3071	3095	3119
3024	3048	3072	3096	3120

Name: 3020   LANCASHIRE FUSILIER

6.6. LONDON TRAMLINK

This system runs through central Croydon via a one-way loop, with lines radiating out to Wimbledon, New Addington and Beckenham Junction/Elmers End, the total route mileage being 18½ miles. It opened in 2000 and is now operated by Transport for London. Six new Stadler trams entered traffic in spring 2012 and four more are on order.

System: 750 V DC overhead.   **Platform Height:** 350 mm.
Depot & Workshops: Therapia Lane, Croydon.

Livery: Light grey & lime green with a blue solebar.

Advertising liveries:

2531 – McMillan Williams Solicitors (black).
2534 – McMillan Williams Solicitors (white & red).

BOMBARDIER   3-SECTION TRAMS

Built: 1998–99 by Bombardier, Vienna, Austria.
Wheel Arrangement: Bo-2-Bo.   **Traction Motors:** 4 x 120 kW.
Dimensions: 30.1 x 2.65 m.   **Seats:** 70.
Doors: Sliding plug.   **Couplers:** Scharfenberg.
Weight: 36.3t.   **Maximum Speed:** 50 mph.
Braking: Disc, regenerative and magnetic track.

2530	2534 **AL**	2538	2542	2546	2550
2531	2535	2539	2543	2547	2551
2532	2536	2540	2544	2548	2552
2533	2537	2541	2545	2549	2553

Name: 2535 STEPHEN PARASCANDOLO 1980–2007

STADLER   5-SECTION TRAMS

Six new Variobahn trams entered traffic in 2012.
Built: 2011–12 by Stadler, Berlin, Germany.
Wheel Arrangement: **Traction Motors:** 8 x 45 kW.
Dimensions: 32.4 x 2.65 m.   **Seats:** 74.
Doors: Sliding plug.   **Couplers:** Albert.
Weight: 41.5 t.   **Maximum Speed:** 50 mph.
Braking: Disc, regenerative and magnetic track.

Advertising livery: 2554 – Love Croydon (purple & blue)

2554 **AL**	2555	2556	2557	2558	2559

On order and due for delivery 2015–16:

2560	2561	2562	2563

6.7. NOTTINGHAM EXPRESS TRANSIT

This light rail system opened in 2004. Line 1 runs for 8¾ miles from Station Street, Nottingham (alongside Nottingham station) to Hucknall, including a short spur to Phoenix Park. There is around three miles of street running through Nottingham. Extensions are under construction to Clifton Lane (Line 2) to the south of Nottingham, and Toton Lane (Chilwell) via Beeston to the west (Line 3) and these are due to open in spring 2015.

22 new Alstom Citadis trams are now being delivered for these extensions. The first of these new trams entered service on Line 1 in July 2014.

The system is operated by the Tramlink Nottingham consortium (formed of Alstom Transport, Keolis, Wellglade, Meridiam Infrastructure, InfraVia and VINCI Construction).

System: 750 V DC overhead.   **Platform Height:** 350 mm.
Depot & Workshops: Wilkinson Street.
Livery: Silver & green with black window surrounds.

BOMBARDIER INCENTRO   5-SECTION TRAMS

Built: 2002–03 by Bombardier, Derby Litchurch Lane Works.
Wheel Arrangement: Bo-2-Bo.   **Traction Motors:** 8 x 45 kW wheelmotors.
Dimensions: 33.0 x 2.4 m   **Seats:** 54 (4).
Doors: Sliding plug.   **Couplers:** Not equipped.
Weight: 36.7 t.   **Maximum Speed:** 50 mph.
Braking: Disc, regenerative and magnetic track for emergency use.

Advertising liveries:

206 – e.on (red).
207 – PayPoint (white & yellow).
209 – Trent Barton Mango tickets (lime green).
211 – Alstom (pale blue).

201		Torvill and Dean	209	AL	Sid Standard
202		DH Lawrence	210		Sir Jesse Boot
203		Bendigo Thompson	211	AL	Robin Hood
204		Erica Beardsmore	212		William Booth
205		Lord Byron	213		Mary Potter
206	AL	Angela Alcock	214		Dennis McCarthy
207	AL	Mavis Worthington	215		Brian Clough
208		Dinah Minton			

ALSTOM CITADIS 402   5-SECTION TRAMS

These cars are currently being delivered and entered service on Line 1 alongside the original cars in summer 2014.

Built: 2013–14 by Alstom, Barcelona, Spain.

Wheel Arrangement: Bo-2-Bo.   **Traction Motors:** 4 x 120 kW.
Dimensions: 32.0 x 2.4 m   **Seats:** 58 (10).
Doors: Sliding plug.   **Couplers:** Not equipped.
Weight: 40.8 t.   **Maximum Speed:** 50 mph.
Braking: Disc, regenerative and magnetic track for emergency use.

216	Julie Poulter	227
217		228
218		229
219		230
220		231
221		232
222		233
223		234
224		235
225		236
226		237

6.8. MIDLAND METRO

Opened in 1999 and operated by Travel West Midlands, Midland Metro consists of a 12½ mile line from Birmingham Snow Hill to Wolverhampton along the old GWR route to Wolverhampton Low Level. Approaching Wolverhampton it leaves the old railway for street-running to the St George's terminus.

An extension is currently under construction from Snow Hill through the centre of Birmingham to New Street station, and this will open in 2015. Future extensions are planned – to Centenary Square and Edgbaston and also eventually to the new Curzon Street HS2 station.

20 CAF trams are currently being delivered to replace the Ansaldo trams and also to serve the New Street extension. There is an option for a further five vehicles.

System: 750 V DC overhead.   **Platform Height:** 350 mm.
Depot & Workshops: Wednesbury.

ANSALDO   2-SECTION TRAMS

All of these trams are due to be stored by early 2015. They will be stored at Long Marston for potential future use if required.

Built: 1998–99 by Ansaldo Transporti, Italy.

Wheel Arrangement: Bo-2-Bo.	**Traction Motors:** 4 x 105 kW.
Dimensions: 24.00 x 2.65 m.	**Seats:** 52.
Doors: Sliding plug.	**Couplers:** Not equipped.
Weight: 35.6 t.	**Maximum Speed:** 43 mph.

Braking: Regenerative, disc and magnetic track.

Standard livery: Dark blue & light grey with a green stripe & red front end.

MW: Network West Midlands silver & pink.
0: Original Birmingham Corporation tram livery (cream & blue).

01, 02, 03 and 13 are stored at Wednesbury depot.

14 has been moved to Long Marston for use in UK Tram development work.

01	(S)		09	**MW**	JEFF ASTLE
02	(S)		10	**MW**	JOHN STANLEY WEBB
03	(S)	RAY LEWIS	11	**0**	THERESA STEWART
04		SIR FRANK WHITTLE	12		
05	**MW**	SISTER DORA	13	(S)	ANTHONY NOLAN
06		ALAN GARNER	14	(S)	JIM EAMES
07	**MW**	BILLY WRIGHT	15		AGENORIA
08		JOSEPH CHAMBERLAIN	16		GERWYN JOHN

CAF URBOS 3   5-SECTION TRAMS

These cars are currently being delivered and the first entered traffic in September 2014.

Built: 2013–14 by CAF, Zaragoza, Spain.

Wheel Arrangement: Bo-2-Bo.	**Traction Motors:** 8 x 65 kW.
Dimensions: 32.96 x 2.65 m.	**Seats:** 52.
Doors: Sliding plug.	**Couplers:** Albert.
Weight: 41.0 t.	**Maximum Speed:** 43 mph.

Braking: Regenerative, disc and magnetic track.

Livery: Network West Midlands silver & pink.

17	21	25	29	33
18	22	26	30	34
19	23	27	31	35
20	24	28	32	36

6.9. SHEFFIELD SUPERTRAM

This system opened in 1994 and has three lines radiating from Sheffield City Centre. These run to Halfway in the south-east, with a spur from Gleadless Townend to Herdings Park, to Middlewood in the north with a spur from Hillsborough to Malin Bridge and to Meadowhall Interchange in the north east, adjacent to the large shopping complex. The total length is 18 miles. The system is a mixture of on-street and segregated running.

The cars are owned by South Yorkshire Light Rail Ltd, a subsidiary of South Yorkshire PTE. The operating company, South Yorkshire Supertram Ltd, is contracted to Stagecoach who operate the system as Stagecoach Supertram.

Because of severe gradients in Sheffield (up to 1 in 10) all axles are powered on the vehicles, which have low-floor outer sections.

System: 750 V DC overhead.   **Platform Height:** 450 mm.
Depot & Workshops: Nunnery.
Standard livery: Stagecoach (All over blue with red & orange ends).
Non-standard/Advertising liveries:
111 and 116 – East Midlands Trains (blue).
120 – Original Sheffield Corporation tram livery (cream & blue).

SIEMENS   3-SECTION TRAMS

Built: 1993–94 by Siemens, Krefeld, Germany.
Wheel Arrangement: B-B-B-B.
Traction Motors: 4 x monomotor drives of 250 kW.
Dimensions: 34.75 x 2.65 m.   **Seats:** 80 (6).
Doors: Sliding plug.   **Couplers:** Not equipped.
Weight: 52t.   **Maximum Speed:** 50 mph.
Braking: Regenerative, disc and emergency track.

101	106	110	114	118	122
102	107	111 **AL**	115	119	123
103	108	112	116 **AL**	120 **0**	124
104	109	113	117	121	125
105					

6.10. TYNE & WEAR METRO

The Tyne & Wear Metro system covers 48 route miles and can be described as the UK's first modern light rail system.

The initial network opened between 1980 and 1984, consisting of a line from South Shields via Gateshead and Newcastle Central to Bank Foot (extended to Newcastle Airport in 1991) and the North Tyneside loop (over former BR lines) serving Tynemouth and Whitley Bay with a terminus at St James. A more recent extension was from Pelaw to Sunderland and South Hylton in 2002, using existing heavy rail infrastructure between Heworth and Sunderland.

The system is owned by Nexus (the Tyne & Wear PTE) and operated by DB Regio.

System: 1500 V DC overhead.   **Depot & Workshops:** South Gosforth.

METRO-CAMMELL   2-SECTION UNITS

Built: 1978–81 by Metropolitan Cammell, Birmingham (Prototype cars 4001 and 4002 were built by Metropolitan Cammell in 1975 and rebuilt 1984–87 by Hunslet TPL, Leeds).

Fleet refurbishment is currently taking place at Wabtec, Doncaster but this will only include 86 cars. The work commenced in 2010 and is due to be completed by late 2015. As part of this work the number of seats is reduced from 68 to 64. Refurbished cars are shown as **TW**.

Wheel Arrangement: B-2-B.
Traction Motors: 2 x Siemens of 187 kW each.
Dimensions: 27.80 x 2.65 m.   **Seats:** 68 (**TW** = 64).
Doors: Sliding plug.   **Couplers:** BSI.
Weight: 39.0 t.   **Maximum Speed:** 50 mph.
Braking: Air/electro magnetic emergency track.

Standard livery: Red & yellow unless otherwise indicated.
B Blue & yellow.
0 (4001) Original 1975 Tyne & Wear Metro livery of yellow & cream.
0 (4027) Original North Eastern Railway style (red & white).
TW New Tyne & Wear Metro (grey, black & yellow).

Advertising liveries:

4002 – Tyne & Wear Metro (orange & black).
4038 and 4084 East Coast trains (silver).
4040 and 4083 – Emirates Airlines (red).
4045 – Newcastle International Airport – 75 years (purple).
4067 – Artist Alexander Millar (white & blue).
4080 – South Shields market (white).

4001	0	4016	B	4031	TW	4046	TW	4061	TW	4076	TW
4002	AL	4017	TW	4032		4047	TW	4062	TW	4077	
4003		4018	TW	4033	TW	4048		4063	TW	4078	TW
4004	TW	4019	TW	4034	TW	4049	TW	4064	TW	4079	TW
4005		4020	TW	4035	TW	4050	TW	4065	TW	4080	AL
4006	TW	4021	TW	4036	TW	4051	TW	4066	TW	4081	TW
4007		4022	TW	4037	TW	4052	TW	4067	AL	4082	TW
4008	TW	4023	TW	4038	AL	4053	TW	4068	TW	4083	AL
4009	TW	4024	TW	4039	TW	4054	TW	4069	TW	4084	AL
4010	TW	4025		4040		4055	TW	4070	TW	4085	
4011	TW	4026		4041	TW	4056	TW	4071	TW	4086	TW
4012		4027	0	4042	TW	4057	TW	4072	TW	4087	TW
4013	TW	4028	TW	4043	TW	4058	TW	4073	TW	4088	TW
4014	TW	4029	TW	4044	TW	4059	TW	4074	TW	4089	TW
4015	TW	4030	TW	4045	AL	4060	TW	4075	TW	4090	TW

Names (to be removed on refurbishment):

4026	George Stephenson	4077	Robert Stephenson

7. CODES

7.1. LIVERY CODES

1 "One" (metallic grey with a broad black bodyside stripe. White National Express/Greater Anglia "interim" stripe as branding).

AL Advertising/promotional livery (see class heading for details).

AN Anglia Railways Class 170s (white & turquoise with blue vignette).

AV Arriva Trains (turquoise blue with white doors and a cream "swish").

AW Arriva Trains Wales/Welsh Government sponsored dark & light blue.

BG BR blue & grey lined out in white.

CL Chiltern Railways Mainline Class 168 (white & silver).

CR Chiltern Railways (blue & white with a red stripe).

EM East Midlands Trains {Connect} (blue with red & orange swish at unit ends).

FB First Group dark blue.

FD First Great Western & First Hull Trains "Dynamic Lines" (dark blue with thin multi-coloured lines on lower bodyside).

FI First Great Western "Local Lines" DMU (varying blue with local visitor attractions applied to the lower bodyside).

FS First Group (indigo blue with pink & white stripes).

FT First TransPennine Express "Dynamic Lines" (varying blue with thin multi-coloured lines on lower bodyside).

G BR Southern Region/BR DMU green.

GA Abellio Greater Anglia (white with red doors & black window surrounds).

GC Grand Central (all over black with an orange stripe).

LM London Midland (white/grey & green with broad black stripe around windows).

LO London Overground (all over white with a blue solebar & black window surrounds).

NO Northern (deep blue, purple & white). Some units have area-specific promotional vinyls (see class headings for details).

O Non-standard (see class heading for details).

RK Railtrack (green & blue).

SN Southern (white & dark green with light green semi-circles at one end of each vehicle. Light grey band at solebar level).

SR ScotRail – Scotland's Railways (dark blue with Scottish Saltire flag & white/light blue flashes).

ST Stagecoach {long-distance stock} (white & dark blue with dark blue window surrounds and red & orange swishes at unit ends).

VT Virgin Trains silver (silver, with black window surrounds, white cantrail stripe and red roof. Red swept down at unit ends).

XC CrossCountry (two-tone silver with deep crimson ends & pink doors).

Y Network Rail yellow.

7.2. OWNER CODES

A	Angel Trains
BB	Balfour Beatty Rail Infrastructure Services
BY	Bridgeway Railcare
CR	The Chiltern Railway Company
CS	Colas Rail
E	Eversholt Rail (UK)
FW	First Great Western (assets of the Greater Western franchise)
H1	Network Rail (High Speed)
HD	Hastings Diesels
MG	Mid Glamorgan County Council
NR	Network Rail
P	Porterbrook Leasing Company
SG	South Glamorgan County Council
SI	Speno International
SK	Swietelsky Babcock Rail
VL	Voyager Leasing (Lloyds Banking Group/Angel Trains)
VO	VolkerRail

7.3. OPERATOR CODES

AW	Arriva Trains Wales
CR	Chiltern Railways
DB	DB Schenker Rail
EM	East Midlands Trains
GA	Abellio Greater Anglia
GC	Grand Central
GW	First Great Western
HD	Hastings Diesels
HT	First Hull Trains
LM	London Midland
LO	London Overground
NO	Northern
SN	Southern
SR	ScotRail
SW	South West Trains
TP	TransPennine Express
VW	Virgin Trains
XC	CrossCountry

7.4. ALLOCATION & LOCATION CODES

Code	Depot	Operator
AK	Ardwick (Manchester)	Siemens
AL	Aylesbury	Chiltern Railways
AN	Allerton (Liverpool)	Northern
CF	Cardiff Canton	Arriva Trains Wales/Colas Rail
CH	Chester	Alstom
CK	Corkerhill (Glasgow)	ScotRail
CZ	Central Rivers (Barton-under-Needwood)	Bombardier Transportation
DY	Derby Etches Park	East Midlands Trains
EX	Exeter	First Great Western
HA	Haymarket (Edinburgh)	ScotRail
HT	Heaton (Newcastle)	Northern
IS	Inverness	ScotRail
MN	Machynlleth	Arriva Trains Wales
NC	Norwich Crown Point	Abellio Greater Anglia
NH	Newton Heath (Manchester)	Northern
NL	Neville Hill (Leeds)	East Midlands Trains/Northern
NM	Nottingham Eastcroft	East Midlands Trains
OO	Old Oak Common HST	First Great Western
PM	St Philip's Marsh (Bristol)	First Great Western
RG	Reading	First Great Western
SA	Salisbury	South West Trains
SE	St Leonards (Hastings)	St Leonards Railway Engineering
SJ*	Stourbridge Junction	Parry People Movers
SU	Selhurst (Croydon)	Southern
TM	Tyseley Locomotive Works	Birmingham Railway Museum
TS	Tyseley (Birmingham)	London Midland
WN	Willesden (London)	London Overground
XW	Crofton (Wakefield)	Bombardier Transportation
ZA	RTC Business Park (Derby)	Railway Vehicle Engineering
ZB	Doncaster Works	Wabtec Rail
ZC	Crewe Works	Bombardier Transportation UK
ZD	Derby Works	Bombardier Transportation UK
ZG	Eastleigh Works	Arlington Fleet Services
ZH	Springburn Depot (Glasgow)	Knorr-Bremse Rail Systems (UK)
ZI	Ilford Works	Bombardier Transportation UK
ZJ	Stoke-on-Trent Works	Axiom Rail (Stoke)
ZK	Kilmarnock Works	Wabtec Rail Scotland
ZN	Wolverton Works	Knorr-Bremse Rail Systems (UK)
ZR	York (Holgate Works)	Network Rail

*= unofficial code.